Modelling the future water infrastructure of cities

Dedicated to the memory of my twin brother
Alexander

MODELLING THE FUTURE WATER INFRASTRUCTURE OF CITIES

DISSERTATION

Submitted in fulfillment of the requirements of
the Board for Doctorates of Delft University of Technology
and of the Academic Board of the UNESCO-IHE Institute for Water Education
for the Degree of DOCTOR
to be defended in public on
Wednesday, September 18, at 12:30 hrs
in Delft, the Netherlands

by

Arlex SANCHEZ TORRES

Master of Science in Water Science and Engineering
specialization in Hydroinformatics, UNESCO-IHE, The Netherlands

born in Cali, Colombia.

CRC Press
Taylor & Francis Group
Boca Raton London New York

CRC Press is an imprint of the
Taylor & Francis Group, an **informa** business
A BALKEMA BOOK

This dissertation has been approved by the supervisors:
Prof. dr.ir. A.E. Mynett
Dr. Z. Vojinovic

Composition of Doctoral Committee:

Chairman	Rector Magnificus Delft University of Technology
Vice-Chairman	Rector UNESCO-IHE
Prof.dr.ir. A.E. Mynett	UNESCO-IHE/ Delft University of Technology (supervisor)
Dr. Z. Vojinovic	UNESCO-IHE (co supervisor)
Em.Prof.dr. R.K. Price	UNESCO-IHE/Delft University of Technology
Prof.dr.ir. L.C. Rietveld	Delft University of Technology
Prof.dr. D. Savic	University of Exeter
Prof.dr. P. O'Kane	University College Cork
Prof.dr.ir. F.H.L.R. Clemens	Delft University of Technology (reserve member)

Published by:
CRC Press/Balkema
Schipholweg 107C, 2316 XC Leiden, The Netherlands

ISBN: 978-1-138-00153-4 (pbk)

Visit the Taylor & Francis Web site at
http://www.taylorandfrancis.com

and the CRC Press Web site at
http://www.crcpress.com

Acknowledgments

I wish to express my sincere gratitude to Professor Roland Price for all his guidance, coaching, support and respect during this research. Thank you for sharing with me your experience and ideas, standing together in front of the blackboard and drawing some points, lines and sketch procedures - that was really fun and I already miss it. Thank you for encouraging me to complete this research and not abandon it. Although you may have found it hard at times to motivate me, you always did. "Thanks for the match" you most of the time carried with you to provide light in my darkness. All this has enhanced my skills to conduct independent research. I should not forget to thank your wife Thea for allowing you to work with me at your home, even after your retirement. Dear Thea, if it can be of any consolation, I think I am the last one.

I wish to thank my supervisors Prof. Arthur Mynett and Dr. Zoran Vojinovic for their advice and patience during this research and for facilitating this learning experience. My sincere gratitude goes to Dr. Zoran Vojinovic with whom I have been working since my master of science topic. You caught my interest and brought to me the idea of starting research in the area of applying agent based modelling theories to urbanization problems. I have learned a lot from you about modelling but also about practical issues in urban hydroinformatics.

My gratitude to Professor Mynett is enormous. Thank you for your willingness to take over from Prof. Price when regulations so required, and for all your support to create the enabling conditions to finalize this research. Your experience proved pivoting to guide me through all practicalities and formalities until the end - and get it done, finally. Thank you for the tremendous energy you put into this process, especially during the most difficult year of my personal life. I much appreciate your understanding and flexibility with the situation.

UNESCO-IHE is a unique learning place where people are transformed since the moment they enter the building. Thank you management team of the institute for providing all the necessary support to conduct this research. Special thanks go to PhD officer Jolanda Boots for all the support during this project.

I also wish to thank Professor Dimitri Solomatine, Dr. Andreja Jonoski, Dr. Ioana Popescu and all my colleagues from the hydroinformatics chair group for sharing their knowledge, experience and for their continued support.

Special thanks go to all my friends at the Institute: Carlos, Leonardo, Gerald, Mijail, Solomon, Nagendra, Girma, Yared, Kiti, Wilmer and many more for their valuable comments, discussions and support.

This study was carried out within the framework of the European research project SWITCH (Sustainable Urban Water Management Improves Tomorrow's City's Health). SWITCH is supported by the European Commission under the 6th Framework Programme and contributes to the thematic priority area of 'Global Change and Ecosystems' [1.1.6.3] Contract n° 018530-2.

I would like to extend my gratitude to Severn Trend as well as to the Birmingham and Belo Horizonte SWITCH learning alliances for allowing me to use their information and data in this study. We are also thankful to Innovyze for providing a research licence of Infoworks CS to UNESCO-IHE. The land use data of Birmingham was obtained from the Corine Dataset of the European Environmental Agency. Additional data for land use was acquired through the Centre for Ecology and Hydrology in the UK (Morton et al., 2011).

Thank you Leonardo for providing me with the dataset for the water distribution case study you used in your MSc study. Coincidentally, there happened to be another available dataset for land use changes over different years. Other sources of information were the municipality's of Villavicencio website, in particular the Plan for Land Use and Development within the set of tutorials that support the ILWIS software. These datasets enabled me to kick off the initial experiments of this research.

I would like to say thanks to RIKS in Maastricht, particularly to Hedwig van Delden and Jasper van Vliet, for their support at the beginning of this research, sharing ideas and providing insight into their modelling tools and knowledge. I am greatly endebted to Deltares, for their financial support during the last stages of this research, in particular to Dr. Frans van der Ven for his valuable feedback and comments about the research findings during our discussions.

During my own research endeavours, I had the opportunity to guide six MSc students, all of whom supported the research in this thesis. The first was Hamisi Matungulu, who continued testing and upgrading the algorithms developed for the rehabilitation of urban drainage networks. The second was Flora Anfarivar who introduced risk into the multiobjective optimization framework developed to rehabilitate drainage networks. The third student was Marwa Waly, who helped me to process the initial dataset acquired for the case study of Birmingham and to test the initial model, ideas and algorithms. The fourth was Neiler Medina who helped me organize and process a large dataset for the case study of Birmingham, and further test and improve the algorithm to derive the network layout. Both contributed directly to chapters 3, 6 and 7. The fifth student was Diego Paredes who tested and enhanced the multiobjective optimization framework for drainage rehabilitation by using a 1D-2D coupled model of SWMM in Quito, Ecuador. The sixth student was Alejandro Corea who started to dynamically model BMP alternatives within SWMM to alleviate flooding as well as control pollution. Thank you all for sharing with me your time and efforts and for posting questions that required not only attention but kept me busy and motivated to help answer them.

Last but not least I would like to thank my family, father, mother, brothers and sisters for their continued support and love from distant Colombia: we grew closer together in difficult times. Thanks to Wim and Maribel in Delft for all your love and support in taking care of Ailèn and David whenever it was important for me to focus on writing down the many parts of this thesis manuscript. I have no words to express my gratitude to my wife Nathasja, for her love and unlimited support during all this time. Thanks for our children Ailèn and David who were born within the framework of this PhD research and for whom I wish a beautiful future on this urbanized planet.

Summary

More than half the world population is living in urban areas, and this trend is likely to continue during the coming decades. As a consequence, many cities around the world are facing considerable pressure to cope with urban development, sustaining economic growth, and providing basic needs and living conditions. In many parts of the world urban infrastructure is aging, in other parts there is uncontrolled urbanization with considerable pressure on economic resources. There is a clear need to be able to predict urban growth and assess the implications for investments and improve the effectiveness of interventions in urban water systems.

It is acknowledged that the interaction between the different subsystems that make up a city is complex. Often the relationship between one system and another is not obvious. Also, the result of certain actions in one part of the system can produce unforeseen consequences in another part of the system or even in a different sub-system, and the relationships between these are not yet well described and understood. This is one of the main arguments for developing integrated tools that can help to advance our understanding of the complex phenomena in urban dynamics.

On the global scale, 95% of urban development seems to occur without proper planning. In some sense, cities can be considered as complex dynamic systems exhibiting characteristics of emergence, self-similarity, self-organization and non-linear behaviour of land use change. All over the world, large scale urban patterns usually arise as a result of interactions between a large number of smaller scale processes that somehow, when combined, create surprising large-scale patterns. The use of tools that can help understand these complexities is important to gain knowledge about the patterns and mechanisms behind urban dynamics. Agent-based modelling is being explored in this thesis, since these techniques allow the representation of the environment (in either two or three dimensions), the integration within GIS, the interaction between temporal–spatial variables, and the interaction between agents and their environment.

This thesis considers the integration of agent-based concepts with physically based hydraulic models of water networks to determine the water infrastructure and performance in delivering adequate water services in the future and how this can shape the urban development process. The objective is to design water systems (water distribution and drainage networks) in the urbanising areas of a city based on the characteristics of the existing networks, and to rehabilitate the system so that it is sustainable. New tools were developed to test this approach to derive the future networks layout. The result is a new approach to urban water infrastructure planning which can help water companies and municipalities to improve the effectiveness of their investments and to be more environmentally efficient.

Along the lines described above, this research covers the use and development of tools and methods to model the future infrastructure needs of cities (based on an analysis of past and current developments), in particular the water distribution and urban drainage networks. The modelling paradigm of Cellular Automata (CA) is used in this thesis to explore scenarios of potential future urban development, land use change, and implications for water management.

For urban drainage in particular, the combination of cellular automata models for land use change with spatial data analysis and urban drainage network models is seen to hold promising results. The research has shown that by analyzing the spatial relation between the drainage network, the road network and the land use, knowledge about the positioning of the main drain conduits can be derived. This yields a new approach to derive the layout of drainage networks of existing systems that can be used to asses scenarios of investment and rehabilitation. Moreover, the approach can be applied to develop case studies in any city on the planet with information currently available on the internet. Any particular case study can be optimized against performance objectives and therefore be compared to the existing system. This approach can lead to indicators that can provide decision makers with information about the level of sub-optimality of the existing system and the required investments to upgrade its capacity.

The developed approach has also been tested to construct future scenarios of urban development that contain the possibility of deriving future network layout. This approach can be optimized and sized to connect to the existing model. This allows the evaluation of the impact of future developments. The three cases studied here were: the city of Birmingham in the UK, Villavicencio in Colombia and Belo Horizonte in Brazil. In doing so, critical elements of the network could be identified and rehabilitation strategies tested in advance.

Clearly there are still limitations of this method, e.g. the availability and type of data required, the quality of the data, all affect the applicability of this approach. But the initial results look promising and the system can easily be expanded when more data becomes available, including information from climate change scenarios.

Samenvatting

Meer dan de helft van de wereldbevolking leeft in stedelijke gebieden en dit neemt in de toekomst waarschijnlijk alleen nog maar toe. Gevolg daarvan is dat veel steden moeite hebben met het realiseren van duurzame economische ontwikkelling en het verschaffen van de benodigde basisvoorzieningen. Op veel plaatsen in de wereld is de stedelijke infrastructuur sterk verouderd, op andere plaatsen breiden steden zich op ongecontroleerde manier uit, met alle gevolgen voor schaarse economische middelen. Er bestaat heel duidelijk behoefte om de groei van steden te kunnen voorspellen en de gevolgen voor de vereiste investeringen en effectiviteit van maatregelen te kunnen beoordelen.

Algemeen wordt erkend dat de interacties tussen deelsystemen in stedelijke gebieden complex zijn. Vaak zijn de onderlinge relaties niet bekend of kunnen de gevolgen van ingrepen niet gemakkelijk worden overzien. Dit leidt tot de noodzaak om te kunnen beschikken over geintegreerde modelsystemen waarmee een beter begrip voor complexe dynamische processen in stedelijke gebieden kan worden verkregen.

Wereldwijd wordt aangenomen dat 95% van de stedelijke gebieden zich uitbreidt zonder enige vorm van planning. Dit betekent dat steden kunnen worden gezien als complexe niet-lineaire dynamische systemen die zich volgens interne wetmatigheden ontwikkelen. Lokale interacties leiden tot globale veranderingen die soms verrassende vormen aannemen. Het gebruik van modellen die deze ontwikkelingen kunnen voorspellen is dan ook van groot belang. In dit proefschrift worden de mogelijkheden van "agent-based modelling" onderzocht (in een omgeving van twee- en drie dimensionale geografische informatiesystemen) om de interactie tussen grootheden in de tijd-ruimte dimensie te onderzoeken.

De mogelijkheden van agent-based modelling worden in dit onderzoek gecombineerd met de meer traditionele aanpak op basis van fysische modellen voor water distributie netwerken om op die manier de behoefte aan water infrastructuur bij stedelijke uitbreiding te kunnen bepalen. Het uiteidelijke doel is om een duurzame aanpak te ontwikkelen voor beheer en onderhoud van water netwerken (zowel aanvoer als afvoer) in stedelijke gebieden.

In dit proefschrift zijn daartoe nieuwe methoden en technieken ontwikkeld en getest met als eindresultaat een aanpak inclusief ondersteunende modelsystemen die van nut kunnen zijn voor water distributiebedrijven en gemeenten en andere overheden, om het rendement van hun investeringen te kunnen verbeteren op een milieuvriendelijke manier.

De ondersteunende modelsystemen bevatten ondermeer methoden om de groei van stedelijke gebieden te kunnen nagaan, waarvoor hier het concept "Cellulaire Automata (CA)" gebruikt is. Met behulp hiervan kunnen ontwikkelingen uit he verleden worden doorvertaald naar mogelijke veranderingen in de toekomst. Aan de hand daarvan kan de vraag naar voorzieningen op het gebied van water aan- en afvoersystemen worden nagegaan.

Water afvoersystemen zijn cruciaal om overstromingen te voorkomen. In dit promotie onderzoek zijn CA-modellen voor veranderingen in grondgebruik gekoppeld met neerslag-afvoermodellen voor regenval, hetgeen tot veelbelovende resultaten heeft geleid. Het blijkt mogelijk om de afvoersystemen zodanig te ontwerpen dat deze aansluiten bij het (veranderde) grondgebruik en wegenstelsel. Deze nieuwe aanpak kan vervolgens worden gebruikt om de vereiste investeringen in aanleg en onderhoud af te schatten. Bovendien kan deze aanpak worden toegepast op vrijwel elke stad waarvan de gegevens op internet beschikbaar zijn.

Elk scenario kan worden geoptimaliseerd aan de hand van vooraf vastgestelde criteria en worden vergeleken met het bestaande systeem. Dit leidt tot een rangschikking van opties op basis waarvan beleidsmakers vervolgens hun investeringsbeslissingen en onderhoudscenario's kunnen bepalen. Deze aanpak is toegepast op verschillende steden (Birmingham in de UK, Villavicencio in Colombia, en Belo Horizonte in Brazilie) met als gemeenschappelijk resultaat dat het mogelijk bleek om kritieke onderdelen in waternetwerken te identificeren en rehabilitatie programma's op te stellen die vooraf kunnen worden getoetst op hun effectiviteit.

Uiteraard kent ook deze aanpak grenzen, bijv. de beschikbaarheid en kwaliteit van de vereiste informatie, maar de eerste resultaten van deze aanpak lijken veelbelovend en de methodiek kan gemakkelijk worden uitgebreid zodra nieuwe gegevens beschikbaar komen, bijvoorbeeld van te verwachte klimaatveranderingen.

Contents

1 Introduction

1.1 Background

The global population keeps growing and as of 2008, for the first time in human history, more than half of the world's population is living in urban areas (UN-Habitat, 2008). Projections suggest that over the next 30 years, virtually all of the world's population growth will occur in the urban areas of low- and middle-income countries, mainly in the South (Garau et al., 2005, UN-Habitat, 2008).

Unplanned settlements are one of the outcomes of this urbanization process. Although in developing countries urbanization is often associated with an increase in a nation's wealth, it is also associated with an increase in squatters and slums which are lacking minimum living conditions. In many cases the informal city or city dwellers are considered illegal settlements and as such are not recognized by governments. Therefore, the provision of basic services to informal settlements is poor; it usually does not include suitable –or indeed any– provision for services such as water supply, sanitation, garbage disposal, roads, storm water drainage, electricity, public transport, schools and health centers.

Huge investments are needed to improve the situation of many urban areas around the world whilst at the same time ensuring environmental sustainability. With increasing population and the uncertainties of climate change in mind, many cities are already planning or executing public works to upgrade the provision of basic services and maintain the levels of service.

The biggest capital investment and expenditure in a municipality is maintaining and upgrading it's water-related networks, which require extensive resources and planning. This implies considerable investments and interventions that may only be feasible once in a certain number of years or even decades, and need to be well planned and executed. The main challenges that cities are facing now and in the near future are the following:

1. Population growth and urbanization
2. Adaptation to climate change
3. Deterioration of infrastructure systems
4. Developing proper governance and policies
5. Absorbing new technologies
6. Saving on energy costs

Urban drainage and water distribution networks are expensive technologies, which aim to transport water to households and discharge wastewater, sometimes in combination with storm water run-off, in such a way that public health is protected and urban flooding risks are reduced.

Traditionally urban planning and development have been based on the formulation and execution of master plans for a fixed period of time. These plans normally include projections of population growth, demand for future services, land use changes, etc. Master plans normally end up on the bookshelf; they are increasingly ineffective as very often these plans are not fulfilled. As a result the goals are lost, and some areas and sectors inside the cities are developed without any control. This is particularly interesting in urban areas in developing countries for several reasons including unexpected events such as migration, natural disasters, economic factors, etc. Since most of the world's population growth will occur in megacities in the developing world, it is estimated that a big proportion of this urban developments will be unplanned.

Since financial resources are scarce, there is an urgent need to optimize them. Increasing the effectiveness of the implemented solutions by better planning and decision making is also needed. To achieve the combined goals of improved efficiency and effectiveness, there is a need to perform integrated analyses. Computer models are generally accepted tools in such optimization processes.

Urban systems have increased in complexity as never seen before in history. Cities evolve based on the characteristics of emergence, self-similarity, self-organization and non-linear behaviour of land use changes with time; see Batty and Langley, (1994). The use of tools that can help in understanding these characteristics are important to gain knowledge about the patterns and mechanisms behind urban dynamics. Agent-based models have been developed to represent evolutionary phenomena in several disciplines of science. For modelling land use changes and urban planning, some interesting aspects used in this technique are the representation of the environment which can be done in two or three dimensions, the integration with GIS, coupling of temporal–spatial variables, the interaction between agents, and with their surrounding environment.

Integrated urban water management is a challenging issue that aims at the sustainable use of water resources so that the demands can be met now and in the future in terms of quality and quantity. The current practices in the sector are leading towards a crisis that is calling for innovative thinking and the adoption of new strategies including integrated thinking and planning. This may sound nice, but the truth in many situations is that the institutional arrangements are so rigid and fragmented that this is not possible. Hence there is a clear need to develop support tools that allow such integration. This research tries to contribute to developing computer-based tools that can help planners and decision-makers at the city level to understand the main drivers affecting the urban water cycle, to analyze future scenarios of city expansion and to anticipate bottlenecks and develop possible solutions. The tool can be used for exploring measures for urban rehabilitation, and for developing planning strategies.

1.1.1 Urban growth and pressure over water systems infrastructure

The world's urban population is projected to grow by more than two billion by 2030, (UN-HABITAT 2003). 94% of this urban population growth will be in less developed regions, and by 2030 the urban population will have, by far, surpassed the rural population. This means that virtually all the additional needs of the world's future population will have to be addressed in the urban areas of low- and middle-income countries.

The development of any urban area within a catchment generates several impacts on the environment and the natural water cycle such as:

- *Growing demand for water*

An unprecedented growth of the urban population is a major driver for urban water management, especially in the developing world. Growth rates of up to 4% per year face cities in developing countries with almost impossible challenges. Planning the city's expansion, providing shelter, energy, water, food, sanitation, health, etc is needed every year for large numbers of people that are the equivalent of the population of large towns. Increased urban water demand may lead to large infrastructural works to transport water over longer and longer distances.

- *Generation of wastewater*

The amount of water that is supplied to the households and other users within the city is converted into wastewater. The more water is supplied, the larger the wastewater flows. The characteristics of the pollution, both in terms of load and quality, depends on the uses, e.g. the (type of) industry, irrigation, domestic, commercial. Wastewater is transported away from houses and buildings via pipe networks to minimize human contact with excreta and pathogens. There are hundreds of substances and toxins that are discharged into sewers during a normal day. The wastewater treatment plants are used to reduce the load of pollutants that are finally discharged to the receiving water body and ecosystem.

- *Alteration of the natural hydrodynamic pattern in the water sources, depletion of groundwater levels.*

By extracting water, the normal dynamics of water flow in the ecosystem nearby the city area is affected, including the water table in the aquifer beneath the city. Quite often the rivers and aquifers are polluted by the wastewater affecting downstream users. Groundwater table lowering due to over abstraction is already a reality in many cities.

- *Increase in impervious and hard surfaces: changing the runoff velocity and collection of dust and solid waste.*

These include the pavement of roads and streets, rooftops, etc. Such impervious surfaces cut off the amount of water that infiltrates. This has a direct effect on the recharge of aquifers and affects base flows in the streams. The increase of hard surfaces increases the surface runoff, and peak flows are larger and water moves faster over these surfaces; therefore the peak arrives earlier and the magnitude of urban floods can be increased.

The impervious areas collect and accumulate dust, all kinds of solids and wastes, pollutants such as those leaked from vehicles, and particles from tires etc. All these substances and particles get diluted and washed off during rainfall, creating a flush flood (like a toilet) that literally washes the streets generating potentially a highly polluted discharge to the receiving water bodies. This urban runoff from storm water is rarely treated and/or even recognized as a problem in developing countries. Sweeping and street cleaning is a major factor to help limit pollution from urban runoff, as well as safe disposal of batteries and containers of toxics and chemical substances such as oil, liquids for car cleaning and maintenance; and insecticides and pesticides commonly used at home and for gardening.

- *Impact on the receiving water bodies.*

The effect of wastewater discharges from urban areas into a receiving water body is difficult to quantify and regulate because of their intermittent and varied nature. Urban storm water runoff always contains various pollutants. Depending on the pollutant's characteristics, different types of damage can result to either aquatic life or people. Many of these pollutant loadings are watershed specific, and vary as the watershed characteristics changes (i.e. street cleaning frequency, traffic load, etc.). The accuracy of drainage water quality modeling is highly dependent on the availability of local monitoring data and the effectiveness in transferring literature values for parameters to a local area (Zhang et al., 2006, Ahlman, 2006). Very often the collection of data on pollutant urban runoff are not included in water quality monitoring campaigns, and data on pollutant buildup on catchments surfaces are extremely lacking in the tropics (Rahmat et al., 2006).

- *Climate change*

Climate change is a critical issue everywhere in the world, and it poses a challenge to humanity to better use the scarce resources that are still available. Some places need to plan and be prepared to re-use water and deal with droughts, and others need to be prepared for flooding and excess. In general, there is a need to use water wisely and to be more efficient. The combination of social growth and urban drainage services provision poses a challenge of optimization. Climate change is an important driver that affects the pressure on the state of the urban water system. Changes in precipitation patterns towards more intense storms lead to an increased risk of flooding. Cities located in urbanized river basins may need to compete with agriculture for water allocations during dry periods.

- *Deterioration of infrastructure systems*

In those cities where a major water infrastructure was put in place during the previous century, urban water managers will increasingly be confronted with deterioration of infrastructure, especially pipe networks. In many parts of Europe, pipes are over 100 years old and the cost of rehabilitation of water infrastructure system is increasing substantially. European cities are spending the order of 5-billion Euros per year for wastewater network rehabilitation (Vahala, 2004). The amount spent on asset rehabilitation programmes will further increase over the coming decades due to the synergetic effects of infrastructure ageing, urbanization and climate change. Infrastructure deterioration will impact on public health, the environment, and institutions in various ways. Higher rates of water leakage mean higher water losses and higher chances of in-filtration and ex-filtration of water. This will create higher chances of drinking water contamination and the outbreak of water-borne disease.

Although the effects of urbanization can be severe, there is a need to explore different options that can help us to minimize these impacts. This process will drive the generation of innovative ideas and opportunities to change the actual trend. For example source control: minimizing the use of water (quantity) by being more efficient (on site sanitation, dry toilets and urinals), good cleaning services (solid waste disposals) and the recognition of the importance for better understanding of the urban water cycle, in particular the interaction between the sub-systems to acquire a holistic view.

1.2 Water distribution problems, planning and design

A water distribution system consists of the catchment where the source of water is located, the source itself (river, aquifer, lake, etc), the water treatment plant, storage tanks, pumping stations, pipes, valves, etc. The water demand depends on living standards, weather, habits, culture, etc.

Traditionally the planning phase of water distribution networks involves the consultation of several stakeholders and authorities at municipal and regional level. Based on those socio-economic plans specific information about the water supply systems can be obtained. Information about the spatial location of the future development, possible water sources locations and the demand for water can be estimated base on the population estimates, housing and industrial, commercial plans. Due to the uncertainty in many of the factors affecting the future development is a common practice to develop water distribution facilities in stages or master plans. This practice provides the opportunities to assess and adapt the design of the expansion of the system in case it deviates from the original ideas.

The design of water distribution systems required the consideration of hydraulic and engineering criteria. The hydraulic performance is assessed in term of the provision of the required water demands, pressure and velocities. It also must ensure adequate functioning during emergency events (fire, pipe burst, etc) and keep the operational cost low. The engineering criteria are also important to ensure the durability of the system during the life expectancy of the several components, the selection of pipe materials, valves, pumps and construction material of other components such as tanks or reservoir are important.

The hydraulic design requires detail calculations because the performance of each component affects the operation on each other. The layout of the system is the first step of the design and it directly affects the costs, the performance of the systems and the operation and maintenance. It is normal to have loop networks in urban areas than in more regional systems. Once the layout is defined the sizing of the system and different elements can be considered as an optimization problem.

The main problems related to water distribution are the ageing of the systems that causes pipe bursts and leakages, the growing demand for water due to population growth, urbanization and economic development. Climate change is causing disruptions in the natural availability of water resources in different places, and therefore demands cannot always be met and water may need to be transported over large distances.

1.3 Urban drainage system problems, planning and design

The drainage and sanitation system of urban areas includes the generation and transport of solid waste, excreta and grey water, as well as storm water drainage. In general the production of waste depends mainly on standards of living, population densities, habits and the characteristics of the water supply services. Storm water depends on climate, meteorology and geology.

The development of an urban drainage network requires large investment by the community. Among the many factors that affect construction and operational costs are the diameters, installation depths, slopes, construction and operation of overflow structure and the use of pumping stations. As a basic principle, urban drainage networks are designed to follow the slopes of the natural terrain to make best use of gravity, and to minimize excavation costs and the use of lifting stations. The combination of these variables, the constraints imposed by the topography and the size of the system make it hard to analyse it manually and computational tools are therefore required. The layout of a drainage network is required in order to size pipes and ancillary structures. The layout of the drainage network depends on the spatial distribution of the land use and it's greatly influenced by the topography, the natural streams and the road network.

Urban drainage is a vital component of urban infrastructure and requires huge investment for planning, design, construction, operation and maintenance throughout the design period. Safe and efficient drainage systems are important to maintain public health and safety, due to the potential impact of flooding on life and property and to protect the receiving water environment. (Vojinovic, 2005).

The population growth and the urbanization process are causing a change in the natural hydrological cycle in any urban area. This is the result of a change in the surface terrain cover and the use of impermeable materials that allow less infiltration, less recharge of aquifers and the generation of faster urban runoff. Climate change is causing a disruption in the natural distribution of rainfall and in some areas of the planet higher rainfall intensities are leading to higher frequencies of urban floods with high economical and social damages and losses.

1.4 Integrated urban water systems analysis and planning

Traditionally the components of the urban water cycle such as water supply, wastewater transportation and treatment, stormwater collection and disposal, have been considered separately for their operation and institutional management. This approach has lead to ineffective planning and delivery of water related services with limited reference to one and other. This has caused an increasing impact on the surrounding environment and water bodies, with the subsequent socio-economic and ecological conflicts. Integrated Urban Water Management (IUWM) on the other hand is an emerging concept that refers to the process of managing freshwater bodies, water supply, wastewater and stormwater as links within the same resources management structure, considering the urban areas as the unit of analysis.

The IUWM approach has emerged from the recognition that an integrated approach in urban water management offers good opportunities for decision making and concrete action. Besides that, it offers a framework to recognize and analyze the effects downstream or upstream of certain actions in other components of the water system (Mitchell, 2004). The main principles for IUWM are summarized as follows:

1. Consider all parts of the water cycle, natural and constructed, surface and sub-surface, recognizing them as an integrated system
2. Consider all requirements for water, both anthropogenic and ecological
3. Consider the local context, accounting for environmental, social, cultural and economic perspectives
4. Include all stakeholders in the process
5. Strive for sustainability, balancing environmental, social and economic needs in the short, medium and long term

In line with the principles described above there are a broad of tools and practices that are employed to make IUWM a reality. Some of them are: water conservation and efficiency; water sensitive planning and design, including urban layout and landscaping; utilization of non-conventional water sources including roof runoff, stormwater, greywater and wastewater; the application of fit-for-purpose principles; stormwater and wastewater source control and pollution prevention; stormwater flow and quality management; the use of mixtures of soft (ecological) and hard (infrastructure) technologies; and non-structural tools such as education, pricing incentives, regulations and restriction regimes (Mitchell, 2004).

The urban challenge dictates a much broader and more ambitious approach than the reduction of poverty and environmental sustainability. It also calls for improved urban planning and design, and the provision of adequate alternatives, innovative thinking and decision making, which respond to the informal urban context. To achieve the combined goals of improved efficiency and effectiveness, there is a need to perform integrated analyses. Computer models are generally accepted tools in such processes.

Strategic plans for the urban water system are often formulated for a long term perspective (15-40 years) because the life cycle of part of the infrastructure is 40 years or longer and because the changes and pressures also develop over this period of time. Some changes occur gradually, but some other changes may have the character of step-changes. The plan needs to take into account the uncertainty around the changes, and therefore needs to be built on a flexible strategy, using technologies and methods that are flexible and that can be applied under different future scenarios.

Projecting and simulating the land use changes in space and time is crucial for the understanding and assessment of consequent environmental impacts. The simulation of human-influenced landscapes changes following different scenarios is helpful to reveal strategy policies that can be modified to improve environmental issues in the future

1.5 Problem Statement and objectives of the research

Cities need to achieve a level of sustainability in their water systems to cope with urbanization and external treats. To do that, there is a need to implement integrated approaches and improve urban planning and decision making. Since the 90's the concept of integrated water resources management has been promoted. The complexity of the water systems and the interlinking between the water sub-systems cause that the actions taken in one part of the system are reflected elsewhere, most of the time difficult to foresee what will be the impacts in a very fragmentally managed sector.

Frequently modelers are asked to predict the complex interactions and links between the management actions or projects developments with the response of the systems. Most of the time to understand the complexity of the multi-casual network of interactions several model tools (physically base and/or data driven) are used. One of the biggest challenges is to integrate the outputs of the different models to understand the system dynamics, the effect of the actions, improve the decision process and clearly communicate with the public.

The use of tools that can help in the understanding of the above–mentioned characteristics are important to gain knowledge about the patterns and mechanisms behind urban dynamics. Agent based models are a good modelling paradigm that can help exploring the characteristics of urban growth.

This thesis explores the application of agent-based models to urban water problems in combination with GIS and standard engineering numerical models to show the impact of the urban dynamics in the evolution of the water systems (pipe networks).

The research aims to look at the evolution of water services according to urban development with time. Given a proposed future scenario, can we identify the way that water distribution and drainage services should be extended from existing urban areas to new development areas?

The main hypotheses of this research are:

Hypothesis 1:	By relating the water distribution properties in existing areas to land use characteristics, the revised and extended network serving the existing and newly developed areas can be estimated.
Hypothesis 2:	By relating drainage to topography (stream network) and land use (including major roads), the revised and extended network serving the existing and newly developed areas can be estimated.

The objectives of the research

- To explore the application of agent-based models in urban water problems.
- To apply the concepts and principles of emergence in the development of urban areas.
- To show the impact of the urban dynamics in the evolution of the water systems (pipe networks).

- To demonstrate the functioning and effectiveness of the models for IUWM on two selected demo cities.

The research questions related to these objectives are:

(i) How to use the concepts of emergence and agent based techniques to urban water problems?
(ii) Is it possible to replicate the land use changes that are observed in reality by applying agent based methods? Are there differences between developed and developing countries?
(iii)Given a certain scenario of urban growth is it possible to identify the way to extend the water distribution network from the existing system to the new developments?
(iv)Given a certain future scenario of urban growth is it possible to identify the way to extend the urban drainage network from the existing system to the new developments?

1.6 Outline of the Thesis

Chapter one contains the introduction to the study, briefly explains the background and magnitude of the problem, the challenge and the objectives of the research.

Chapter two contains the literature review that highlights basic principles of urbanization and urban growth models and theory, water distribution networks planning and design, theory and models to design the system, urban drainage and sewer modeling, models available for urban drainage modeling (SWMM 5.0) genetic algorithms, optimization (NSGA-II). Review previous experiences in the field of optimization applied to urban water systems.

Chapter three describes in detail the urbanization phenomena, the problem and modeling approaches to assess urbanization growth. A review of the models that are available for urban growth and land use change are presented. The description of the model engine that is used and the set up of the models is also presented and discussed here. It describes the methodology applied in this study to assess land use change. It formulates the objective functions, the tools, constraints and algorithms used in this part of the study.

Chapter four describes the methodology, formulates the objective functions, the tools, constraints and algorithms used to assess the connection between land use changes and water distribution network. It also presents the approach to obtain the future layout of the system, the design of the future network and the implications for the existing infrastructure. The approach is tested in a case study are in Villavicencio, Colombia.

Chapter five presents the development and construction of the approach to assess the evolution of urban drainage networks. It describes the tools, models, algorithms and constraints used to connect the changes in land use to predict the future layout of the urban drainage network. The layout of the network is tested in a case study area in Birmingham, UK. The approach to derive the drainage network is tested in the existing area where the existing layout is known. The model is used in conjunction with the land use change model to assess future scenarios of urban growth and the consequences of that growth in the existing drainage network.

Chapter six addresses a generalization of the methodology to develop models for an urban area, based on the available data sets on the internet. The approach enables the user and decision makers to develop prototype tools to assess the consequences of urban growth on the water infrastructure. The proposed approach makes use of the information and rules found in the cases of Villavicencio and Birmingham to formulate and urban growth model for the study area of Belo Horizonte. The result of the land use change model for the proposed scenario for the year 2037 is used to derive a future drainage network layout for the Belo Horizonte area and the possible effects of this growth are assessed in a small catchment in the area of Venda nova.

Chapter seven summarizes the findings of the research in the form of conclusions and recommendations.

2 Strategic Planning for Integrated Urban Water Management

2.1 Urban Planning and Growth

There are several approaches for urban development modelling. One of the first models that are described in the literature is Von Thunen model for agricultural allocation. Von Thunnen considered the inter-relation of three factors: the distance of the farmers to the market, the price received for the farmers for of their goods and the land rent cost. The hypothesis was that the intensity of land use was inversely proportional to the distance from the market or the transportation cost. Considering one city as the central market and a flat topography around it, the Von Thunnen model generates a concentric land use pattern with the less intensive land use farthest away from the city center.

With the development of the digital computer a new era in modelling started, due to the ability to handle complex mathematical formulations. The new computational capacity enabled the construction of several models mostly, transportation models, economic, land use allocation, etc. The developments of these models used a wide number of techniques like linear analysis, mathematical programming, simulations etc.

The development of geographical information systems (GIS) and the integration of GIS with urban models have enrich urban development modelling by providing more data sources and new techniques to handle data and present outcomes. These developments have contributed to understand cities as evolutionary and complex systems.

2.1.1 Ecological approach

This approach is based on the belief that human behaviour is determined by ecological principles, such as competition, selection, succession, and dominance. This was started at the Chicago School of Human Ecology in the 1920s, and the most notable models of this approach were Burgess's (1925) concentric zone model, Hoyt's (1939) sector model, and Harris and Ullman's (1945) multiple nuclei model (Liu, 2009).

Burgess's model of urban growth was based on the notion that various elements of a heterogeneous and economically complex urban society actively compete for favourable locations within the city.

Hoyt (1939) developed the sector model in which he identified that homogeneous areas of residence tended to grow outward from the centre toward the periphery in wedge-shaped sectors. In his sector model, in addition to the obvious emphasis on transportation routes where urban growth was often focused on, Hoyt also considered the effects of topographic variations and the adjacent and nearby land use on urban development.

In the Harris and Ullman (1945) model, the patterns of urban growth and change still followed the general ecological principles identified by Burgess. For example, some activities always tend to be located in the vicinity of each other, and others repel each other, whereas some cannot afford the high rents demanded for the best sites. However, this growth was not centered around one single central business district but on certain growing points or "nuclei."

2.1.2 Socio-physical approach

The social physical approach was based on the concept of human interaction in space. It uses an analogy to physics. That is, it uses Newton's Law of Gravitation as an analogue for social interaction between places. It proposed that the movement of human activities such as changes in residence and employment between places were directly proportional to the mass of the activity at the origin and destination, and inversely proportional to the cost (in terms of distance or time) separating them. The model developed from this analogy was referred to as the gravity model, which was widely applied in studies of migrations, settlement network, and the intra-urban structure in the 1960s.

Following the extensive applications of the gravity models in urban spatial interaction studies, Wilson (1970) developed the social physical approach by introducing the second law of thermodynamics—the maximum entropy law—into this approach. Based on the principles of the maximum entropy law, Wilson formulated his entropy-maximizing spatial interaction model. In this model, the movements of people and goods in cities were treated in the manner that particles in gases were treated in statistical mechanics using grand canonical ensembles and distinguishing them by origin and destination as "types" and by origin–destination pairs as "states".

In a typical gravity model, factors such as basic employment, economic structure, and population were usually distributed using particular allocation functions. Models developed under this approach were aggregates; they stressed group behavior rather than individual behaviour.

2.1.3 Neo-classical approach

The neoclassical approach was built on the belief that the process of urban development is essentially an economic phenomenon, being driven by market mechanisms and the natural forces of competition among economic activities and social groups in an urban area. According to the economic theory of equilibrium, the allocation of urban land to various users in both quantitative and locational aspects is controlled by supply-and-demand relationships obeying the general rule of least costs and maximum benefits (Liu, 2009).

2.1.4 Behavioral approach

The central concern of this framework was the behaviour patterns that were the representations of human actions. Urban development was viewed as an end result of human actions, and the value system of urban society as the primary source of the impulse for actions. The objective of this framework was to seek explanations of urban development in terms of human behaviour, with the behavior patterns being a function of people's values. The fourth element of this framework, the control process, concerns how influence could alter or affect behavior patterns and thereby modify urban development toward certain predetermined goals. This element is often referred to as urban development strategies and plans. Under this framework, urban development was first viewed as the consequence of certain strategic decisions that structure the pattern of growth and development, and then as the consequence of the myriad of household, business and government decisions that followed from the first key decisions (Liu, 2009).

2.1.5 Systems approach

All the elements in the system are linked and interrelated and are also linked to the system's environment. For instance, an urban system consists of a set of elements or subsystems, such as population, land, employment, services and transport, to mention a few. All elements within the system are interacting with each other through social, economic, and spatial mechanisms while they are also interacting with elements in the environment (Liu, 2009).

The significance of any one element does not depend on itself but on its relationships with others. It is the links between the different elements of the system that determine its evolution and so permit the process of change in the system. Thus, the focus of the systems approach is not on any single element but the connections and processes that link all the elements. This approach builds on the foundations and concepts developed and presented by Forrester, 1968 principle of systems and later on Forrester, 1969, urban dynamics. In his book urban dynamics, Forrester describes a computer model with hypothetical driving forces that balance population, housing and industrial development.

In order to illustrate the structure and behaviour of systems, a diverse range of mathematical methods has been employed. This includes factor analysis, principal component analysis, multicriteria analysis, linear and nonlinear programming, as well as dynamic systems simulation.

2.1.6 Cities as self organizing systems

Based on the understanding of the open system theory, the process of urban development is being looked at in new ways. A city can be viewed as an open and complex self-organising system that is far from being in equilibrium, and it exists in a constant exchange of goods and energy with other cities and its hinterland. The structure of this system emerges from local actions where uncoordinated local decision making may give rise to coordinated global patterns. Urban development is thus a spatially dynamic process, exhibiting some fundamental features of a self-organising system (Liu, 2009).

This understanding suggests that a ground-up approach under the self-organising paradigm to address the local behaviour of the system is more realistic in modelling urban development, which has resulted in the emergence of a new class of simulation models (Benenson and Torrens 2004; Batty 1997, White and Engelen 1994) geosimulation based on automata, and the agent-based model.

Urban models based on the automata technique have also emerged under the paradigm of a self-organising system, with cellular automata being the simplest but most popular in action. An automaton is an entity that has its own spatial and non-spatial characteristics but also has the mechanism for processing information based on its own characteristics, rules, and external input (Benenson and Torrens 2004).

The multi-agent systems are designed as a collection of interacting autonomous agents, each having its own capacities and goals, but together they relate to a common environment. This type of model operates on the same principles as the cellular automata model, with each agent being considered as individual autonomous agent-automata (Torrens 2003), and their states generally represent some agent-based characteristics. However, distinctions between cellular automata and multi-agent systems exist in a number of ways. One distinction is that in the multi-agent system, the basic unit of activity is the collection of agents representing individuals, developers, planners, or government decision-makers. The agents are autonomous in that they are capable of making independent actions, their activities are directed toward achieving defined tasks or goals, and their influence on the environment can be at different scales.

Another distinction between the cellular automata and the multi-agent systems is that cellular automata are fixed cells in the CA lattice, whereas the agents in the multi-agent systems are dynamic and mobile entities that can move within the spaces that they "inhabit" (Torrens 2003). These agents also can process and transmit information while they move along the spaces and pass the information from one agent and environment to another in their neighbourhood. Consequently, the neighbourhood relationships in agent automata are also dynamic: when individual agents alter their locations in space, their neighbourhood relationships also change.

2.2 Expansion and design of water distribution systems

The growth of the urban population is a major driver for urban water management, especially in the developing world. Growth rates of up to 4% per year face cities in developing countries with almost impossible challenges. Planning the city's expansion, providing shelter, energy, water, food, sanitation, health, etc is needed every year for large numbers of people that are the equivalent of the population of large towns. Increased urban water demand may lead to large infrastructural works to transport water over longer and longer distances. This together with the combine effects of ageing infrastructure and climate change is causing a disruption in the water balance of many urban areas. Due to this, demands cannot be met and water needs to be transported from larges distances, expansions of the existing capacity need to be planned and the operational efficiency and efficient use of water need to be addressed.

The water distribution system consists of the catchment where the source of water is located, the source itself (river, aquifer, lake, etc), the water treatment plant, storage tanks, pumping stations, pipes, valves, etc. The water demand depends on living standards, weather, habits, culture, etc.

The design of water distribution systems required the consideration of hydraulic and engineering criteria. The hydraulic performance is assessed in term of the provision of the required water demands, pressure and velocities. It also must ensure adequate functioning during emergency events (fire, pipe burst, etc) and keep the operational cost low. The engineering criteria are also important to ensure the durability of the system during the life expectancy of the several components, the selection of pipe materials, valves, pumps and construction material of other components such as tanks or reservoir are important.

The hydraulic design requires detail calculations because the performance of each component affects the operation on each other. The layout of the system is the first step of the design and it directly affects the costs, the performance of the systems and the operation and maintenance. It is normal to have loop networks in urban areas than in more regional systems. Once the layout is defined the sizing of the system and different elements can be considered as an optimization problem.

2.2.1 Design of water distribution networks

The goal of the use of optimization for design of WDN's is to obtain a minimum cost of new infrastructure to be deployed while keeping as constrains some measures of state variables of the network like minimum pressures at critical points or supplied demands.

Usually the design is suggested for a single demand or a pattern of demand, since the system is not constructed this is usually assumed by the designer. Mathematically WDN design (or rehabilitation) is an intractable problem (Gupta I. et al., 1993) and its complexity is referred to as NP-hard (Eusuff M.M. and K.E. Lansley, 2003) implying that the solution cannot be found in polynomial time. This means that a rigorous algorithm used for this purpose is not practical and then random search techniques like genetic algorithms are more efficient. The reason resides in the fact that the feasible region of diameters considered as decision variables is non-convex since the constrains are implicit functions of the diameters. The objective function presents multiple local minima and the system solution is based on a nonlinear system of equations. In general, for small cases it is possible to find solutions in short simulation times, but in large networks this is an open and challenging field.

In the case of rehabilitation of a WDN using optimization approaches, the goal is to perform analysis about which part of the network is more suitable of being replaced due to poor service or aging of structures. Usually the objective is to obtain the minimum cost of the replacement of structures along the network (such as pipes, valves and pumps), that corresponds to a minimum investment.

2.2.2 Modelling of water distribution networks

Since water distribution networks design required the postulation of a set of equation that are non linear, the use of computational tools is required. The development of models for water distribution network helps with the understanding of the system and assesses its performance under different circumstances.

Since the birth of electronics, numerical modeling of WDN has been in progress and many algorithms had been implemented. Basically all of them gather from the concept of maintaining a water balance in the nodes and second of preserving energy losses in pipes or loops depending on the method. The first algorithms developed for the simulation of WDN date back to the first half of last century known as loop balance of heads and loop balance of flows (Cross, H. 1936). Then after a while with the growth of the digital computation several other algorithms were born based on Newton's algorithm (Martin D.W. and G. Peters, 1963) and global linearization techniques (Wood D.J and C.O. Charles; 1972) until the development of what we know today as the Global Gradient Algorithm (GGA) developed initially by Todini, E. 1979.

One may want to answer what-if questions about the behaviour of the network under certain conditions (some uncertain) and set some strategic planning. At the same time one may want to be able to guarantee that the supply is performed in an economically way, see for example Savic and Walters,1997, Walski et al 2003. Many measures can be used as operations such as closing and opening valves, changing the pump schedule or testing the behaviour of the network under new scenarios of demand, energy cost patterns or added infrastructure due to urban growth.

2.3 Expansion and rehabilitation of urban drainage systems

Drainage and sanitation of urban areas includes the generation and transport of solid waste, excreta and grey water, and storm water drainage. In general the production of waste depends mainly on living standards, population densities, and the characteristics of water supply services provision. Storm water depends on climate, meteorology and geology.

Urban drainage is a vital component of urban infrastructure and requires huge investment for planning, design, construction, operation and maintenance throughout the design period. Safe and efficient drainage systems are important to maintain public health and safety, due to the potential impact of flooding on life and property and to protect the receiving water environment. Getting reliable data of existing and projected storm water flows is a prerequisite for cost-effective urban drainage design and analysis (Vojinovic, 2005).

Urban drainage systems can be classified according to two types of flow: i.e. wastewater and storm water flow. The relationship between the conveyance of wastewater and storm water has remained complex for urban drainage system management (Price, 2005).

Basically there are two types of conventional sewerage systems:

1. Combined system - in which wastewater and storm water flow come together in the same conveyance system
2. Separate system- in which wastewater and storm water are kept in separate conveyance systems. Usually those separate conveyance pipes lay side-by-side.

Urban drainage as we know them today was one of the outputs of the industrial revolution and its associated urbanization. At the beginning it mainly dealt with the transport of generated volumes of storm water and wastewater. Nowadays, with the increasing complexity of urban areas and all range of waste produced in the cities, water quality and the impact of the discharges from sewer systems in the environment of the receiving water body are equally important to hydraulics and flood management. Urban drainage has become a subject of resource management problem on its own. The appropriate solution involves not only structural but also non-structural aspects, such as planning and operational procedures, environmental impacts and economic and social concerns.

Modern urban drainage systems planning and management cannot even be considered without involving the development and application of mathematical models (Vojinovic, 2005).

There are several aspects that must be considered carefully when dealing with urban drainage problems, namely:

- The hydraulic characteristics are time varying;
- There are human imposed modifications in addition to seasonal changes;
- The flow systems can consist of open channels as well as closed conduits subject to surcharging and backwater effects (providing unsteady and non-uniform flows);
- Spatial and temporal variations of rainfall;
- Water quality and quantity must be considered together;
- Uncontrolled and controlled system discharges because of overflows of either combined or separate sewer overflows

2.3.1 Urban drainage design

Understanding hydraulics is essential in the design of urban drainage systems in order to specify appropriate size of the different components of the system, such as pipes, channels shape, size of storages, etc. It is also needed in the analysis and modelling of existing systems in order to predict the relationship between flow-rate and depth for varying inflows and conditions.

Hydraulics knowledge is also essential to realize the flow characteristics and causes of overflow in a network and to take remedial action for flooding. One of the aspects that require special attention and analysis in urban drainage design is the provision of sufficient conveyance capacity for computed and/or anticipated storm volume or events.

The collection system can be handled either by piped systems or natural or artificial channels. A pipe system can be an open channel or a pressurized system sometimes. Usually storm water collection is performed with gravity except for some flat areas that need pumping systems to enhance the flow. If the hydraulic design of the system is not appropriate for estimated runoff volume, peak flow or collected storm water is not conveyed properly from the surrounded catchment areas, it would cause inconveniency, flood damage, and further health risks.

In sewerage systems there are a variety of flow conditions encountered (Price, 2004), ranging from:

- Free surface flow to surcharged flow (pressurized);
- Steady to unsteady flow;
- Uniform to non-uniform flow (gradually or rapidly varying flow).

Flows in conduits and open channels normally have a uniform cross-section along its length and a uniform gradient. Because the dimensions of the cross section are typically one or two orders of magnitude less than the length of the conduit, unsteady free surface flows can be modeled using one dimensional flow equations.

Flood waves in conduits show certain distinctive characteristics (Price, 2004) including:

- Translation (the peak propagates downstream direction);
- Attenuation (gravity forces tend to flatten the peak or disturbance along the conduit or channel);
- Distortion (change in shape of the wave profile).

This behaviour is well represented by the St Venant equations.

These are presented below for continuity and momentum.

$$\frac{\partial Q}{\partial x}+\frac{\partial A}{\partial t}=0$$ **Continuity** **Eq 1**

$$\frac{\partial Q}{\partial t}+\frac{\partial}{\partial x}\left(\frac{Q^2}{A}\right)+gA\frac{\partial y}{\partial x}-gA(S_0-Sf)=0$$ **Momentum** **Eq 2**

Local acceleration term	Convective acceleration term	Pressure Force term	Gravity Force term	Friction Force term

Inertia Term (Local acceleration term + Convective acceleration term)

Where:

Q	: discharge	m^3/s
y	: water depth	m
A	: cross sectional area	m^2
g	: gravity	m/s^2
So	:bed gradient	
Sf	: friction slope	
t	: time	s
x	:distance	m

There are several commercial software applications that use different numerical schemes to solve these equations simultaneously (SWMM, MOUSE, INFORWORKS etc.).

Transition between free surface flow and pressurized flow and vice versa in a conduit is rather complex. To solve this problem and recognizing the similarity in the governing equations, Preissmann suggested an approach whereby the standard method for free surface flow can be retained even when the flow is pressurized by using a conceptual vertical slot (Preissmann slot) (Price, 2004).

2.3.2 Urban Drainage Rehabilitation and maintenance

Sustainability of urban drainage networks requires continuous rehabilitation and maintenance practices. Rehabilitation in this case is used in its widest sense of maximizing the use of existing assets either by operation or through physical changes, which may include repair, renovation and renewal. The main purpose of the rehabilitation activities is to achieve the desired levels of service, at the minimum cost and with the smallest environmental impacts. The 'level of service' is defined as the standard of performance that is desired. This level of service is the ultimate statement of accountability to homeowners who may get flooded or suffer from other inconveniences.

There are a number of factors that can lead the collapse of sewer system which are constantly subjected to physical, chemical, bio-chemical and biological stresses, within the main factors are:(Stein D., 2001)

- Infiltration / exfiltration: water enters or leaves from the conduits, these leaks can occur in pipe joints, manholes, pipes walling and other structures.
- Flow obstacles: are objects or materials lying in the cross section of the conduit that reduces the hydraulic capacity of it.
- Positional deviations: is the unplanned deviation of sewers and structures due to: hydro-geological changes, load changes, settling, subsidence and earthquakes.
- Mechanical wear: is the removal of the piping material from inner wall surface, which increase the roughness of the walls and reduce the wall thickness and the bearing strength and water tightness.
- Corrosion: is all reactions to non-metallic and metallic materials with their environment, then the consequences could be leakiness, reduction of wall thickness, increase in the roughness of the wall, etc.

- Deformation: the possible consequences of damages are: reduction of hydraulic effectiveness, blockages, increase of maintenance effort, cracks, leaks, pipe break, and collapse.
- Cracks, pipe breaks and collapse: cracks and breaks are originated by the increment of internal and external loading and then collapse that is the complete loss of carrying capacity.
- Development of an urban area: this effect of the human being not only has an impact in increasing the volume and peaks of surface runoff in comparison with infiltration, but also can affect the quality due to the pollutants and sediments that are washed off.
- Lack of capacity of the sewer conduits.
- Blockage in gullies.
- Poor sewer design.
- Aging of its structures.

Since this is an integrated system all these factors are acting continuously on the structures and different elements and all factors can act in conjunction at the same time. The first step in the analysis of the optimum rehabilitation schemes is to have a good knowledge of the current condition and performance of the system.

Butler D. and Davies J., 2000, in the Sewerage Rehabilitation Manual of the UK, recommend some procedures to follow in the evaluation of the performance of the sewerage system, which are divided in phases:

1 Information: review and collect all existing information related with the entire network layout (inventory), besides information of hydrology, hydraulic, environmental, social, structural and operational performance. If additional data is missing survey activities should be performed.
2 Hydraulic assessment: the hydraulic modeling of the system aim to obtain conveyance capacity, flooding frequencies, identify critical zones, and to verify the cost-benefits of rehabilitation measures. Such measures can be applied to maximize the capacity of the system; reduce inflow by diverting them to avoid hydraulic overloading; attenuation of peak flows by using retention and detention ponds, etc.
3 Environmental assessment: Water quality models are used to detect polluted overflows, concentration of pollutants which are detrimental to human consumers and aquatic life.

4 Structural assessment: The evaluation of the physical condition of the asset, which can be use direct and indirect techniques. Direct inspection is made normally by the maintenance crew. The limitations for this method are: the size of the conduits (to enter a person), and presence of poisonous gases. Indirect technologies commonly are carried out by closed-circuit television (CCTV).
5 Social: Surveys to the public play an important role in the rehabilitation, for instance odours, repeatedly events of flooding, blockage in gutters etc.
6 Operation investigation: Focus on the operational and maintenance procedures, which are related with the previous phases that are concerned of the physical performance of the system.
7 Develop solutions: the solutions should take into account hydraulic, environmental problems in an integrated way to be cost effective, consider the structural investigations, and implement an operation and maintenance plan.

2.3.3 Urban Drainage modelling

Urban drainage models aim to understand and predict the behaviour of the components (pipes, channels, culverts, manholes, etc) in such a way that several alternatives (structural and operational changes) combined in scenarios can be tested and evaluated (Price, 2004).

The need for urban drainage and wastewater control and management as a part of complex system requires the use of modelling tools. Traditionally, engineers collected data on elements of an urban drainage system and the environment and then prepared reports on the system performance (Price, 2004). The increasing value and quantity of collected data and the storing requirements resulted in the use of computers. Several approaches exist ranging from physically-based methods, data driven models, hybrid models (conceptual + data driven models, physically base + data driven models), to simple statistical regression models or empirical equations. For example the rational method is a simple model to convert rainfall into runoff that can be used to look at the likely effects of different rainfall intensities.

Computer programs for drainage design and analysis emerged in the 1970s. The model SWMM (Storm Water Management Model) first appeared in the USA in the early 1970s (Rossman, 2005) and has continued to be developed ever since. In the UK in the 1970s, the transport and road research laboratory had developed computer-based hydrograph methods. Though it lacked to develop a standard software package until the early 1980s (Butler and Davies, 2000). Hydroworks and Infoworks are the most recent packages in evolution of initial UK computer-based hydrograph model (Price, 2004).

MOUSE is a popular European drainage model package, developed by DHI - Water and Environment (http://www.dhigroup.com/Software/Urban/MOUSE.aspx). Both SWMM and MOUSE are deterministic models meaning that one combination of input data will always give the same output. In general this commercial software packages have a numerical scheme (engine) to solve the fluid dynamics equations (St Venant equations). Urban drainage models have many uses. Among them the main uses are the design of new systems, the analysis of the performance of existing systems, the evaluation of operational real-time control measures and also for flood forecasting. In design, the physical details of proposed drainage systems are determined so that the system will behave satisfactorily when exposed to specific conditions. In analysis (simulation), the physical details of the system are known and the model-user is interested on how the system responds to specific conditions. i.e. (water depth, pipe flow, discharges, combined overflows, pump rates, surcharge, surface flooding, etc).

2.3.4 Urban flood management

Urban flood management has usually focused on storm water drainage and flood protection with the provision of enough capacity in conduits and channels in the drainage system. The system should accommodate the expected increase in storm water discharge resulting from catchment urbanization. Improving the capacity of the system and upgrading the conveyance capacity is a widely applicable and conventional flood mitigation measure. It includes enlarging (replacing) incapable drainage networks (i.e, widen the channel, or replacing existing pipes with capable pipe sizes), expanding work where lack of drainage network exhibits, detention ponds and others.

The management of storm water systems has different objectives depending on the strategy. In the short term, the priorities are runoff control and pollution mitigation strategies. The medium term objectives focus on development and implementation of water quality improvement, water conservation and strategies to preserve the hydrology and natural pattern of the catchment. The long term objectives place greater emphasis on preservation of natural resources and the amenity value of water in the urban environment for recreational activities and to promote an increased awareness of environmental issues. Although these objectives may initially appear to be somewhat idealistic, especially considering the existing situation in many developing countries, it is important that planners and designers of urban drainage systems aim to satisfy the need of future generations within keeping the objectives of sustainable development as defined by the World Commission on Environment and Development in 1987 (Parkinson J. and Mark O., 2005).

In most cases, the optimal strategy for flood control is one which combines structural measures with nonstructural measures, since, structural flood mitigation works are usually expensive and can create other social problems and inconvenience during the construction period.

Non-structural measures include rainwater harvesting, wetland and flood retarding basin development, preserve and/or increase the pervious characteristic of land, cleaner production principles applied to industry and household level to promote and improve efficient use of water, etc.

2.3.5 Urban flood impacts

Urban flood is a potential risk since the pervious areas and slope of streets and highways increase the velocity of the water and the presence of buildings, bridges, channels and all sort of obstacles can create high water levels in certain areas. Moreover, in tropical countries and the effect of climate change, amongst others heavy rainfall, can create flash floods that leaves little time for warning. Flood damage is proportional to the volume and the velocity of the water (Vojinovic, 2005).

Disasters caused by flooding can cause tremendous tangible but also intangible damage to an urban area and even affect the national economy of a country (Lekuthai, A. and Vongvisessomjai, S., 2001). Tangible damage is the damage that can be readily measured in monetary value, while intangible damage cannot be directly measured in such terms (Kuiper, 1971).

2.4 Modelling tools and paradigms for integrated urban water systems planning

A model is a representation or abstraction of the world we observed. In general terms models can be considered as a device or mechanism that represents a theory that generates information (outputs) from a set of inputs and assumptions. The output of the model helps in the understanding and the adequacy of the theory embedded in the model. In the real world almost every system is complex, and this complexity and dynamics make them hard to understand and analyze. Therefore, it is a common practice to explore those systems using simplified representations, symbols, laws, rules and processes. In this way systems can be explained, analyzed and managed.

The application of models in scientific research fields is very important. Not only do they provide a framework to express and formulate theories, they enable hypothesis and theories embedded in models to be tested and proved. Models play an important role in day to day engineering practice as tools to simulate and predict the behavior of many systems (transport networks, drainage and water distribution networks, aircraft design, etc); also for social systems that are of concern for environmentalist, urban analyst, planners and decision makers. Due to scale, cost and practical reasons is very difficult to manipulate or experiment with full scale systems, therefore researchers construct models to represent the real situation that allow them to explore, understand and predict the behavior of the systems. So, models provide an artificial environment to carry out experiments that otherwise will not be possible. One of the main drivers to construct models is to answer "What if" type of questions to support planners and decision makers.

Models can be classified in different ways; in general there are three groups or categories: 1. Scale models where the reality has been altered by the scale, such as building models used by architects or coastal models to simulate hydraulics and sediment transport at laboratories. 2. Conceptual models which focus in the description of the relations between the different components of reality, for example the integrated urban water management cycle, which describes the links between the different uses and users of water in urban areas. 3. Mathematical models are in the highest level of abstraction, they describe reality by using mathematics, equation, physics laws, etc. For example the formulation of the fluid dynamics equations and methods to solve them are the basis of widely applied models.

Models applied for urban water management are mostly mathematical models. Traditionally, in the urban water sector every component of the water system is modeled individually; even more, the management and institutional arrangement is often fragmented as well. This fragmentation has caused a form of water crisis, since this arrangement makes it difficult for every group of modelers or practitioners to experiment or evaluate the effect of their actions and decision on the different components of the water system either upstream or downstream.

Urban water management needs to be understood as a highly integrated physical and social system. While a lot of effort has been put in understanding the physical part of the system (assets, pipes, etc), little has been done in understanding the social part. Cities are complex systems that are not at equilibrium; their dynamics is related to chaos theory, the principle of emergence and self-similarity as well as fractal geometry.

Recently, there has been progress in the understanding of cities by applying agent based models and cellular automata for land use changes modelling (Engelen et al 2007, Batty et al, 1994, Barredo et al, 2003, van Delden, 2007).

2.4.1 Agent based models

The history of the agent-based model can be traced back to the Von Neumann machine, a theoretical machine capable of reproduction. The device Von Neumann proposed would follow precisely detailed instructions to fashion a copy of itself. The concept was then improved by von Neumann's friend Stanislaw Ulam, also a mathematician; Ulam suggested that the machine be built on paper, as a collection of cells on a grid. The idea intrigued von Neumann, who drew it up—creating the first of the devices later termed cellular automata. von Neumann, 1949.

John Conway formulated the well-known game of life; which unlike von Neumann's machine operated by tremendously simple rules in a virtual world in the form of a 2-dimensional board.

The creation of agent-based models of social systems is often credited to the computer scientist Craig Reynolds. He tried to model the reality of lively biological agents, known as artificial life, a term coined by Christopher Langton.

An agent-based model (ABM) is a computational model for simulating the actions and interactions of a set of individuals (Agents) in a network to asset their effects in the global system behavior. It combines elements of game theory, complex systems, emergence, computational sociology, multi agent systems and evolutionary programming.

The models simulate the simultaneous operations of multiple agents, in an attempt to re-create and predict the actions of complex phenomena. The process is one of emergence from the lower (micro) level of systems to a higher (macro) level. The individual agents are presumed to be acting in what they perceive as their own interests, such as reproduction, economic benefit, or social status (Axtell et all, 2003).
An agent is described as an object with certain characteristics (Tzima et all, 2006):

- autonomous – it operates without the direct intervention of others and has some kind of control over its actions and internal state.
- social – it interacts with other agents using an agent-communication language;

- reactive – it perceives its environment and responds to changes that occur in it;
- proactive – it is able to exhibit goal-directed behaviour.

The three ideas central to agent-based models are social agents as objects, emergence and complexity.

Agent-based models are situated in networks and in lattice-like neighbourhoods. The location of the agents and their responsive and purposeful behaviour are encoded in algorithmic form in computer programs. The modelling process is best described as inductive. The modeler makes those assumptions thought most relevant to the situation at hand and then watches phenomena emerge from the agents' interactions. Sometimes that result is an equilibrium. Sometimes it is an emergent pattern. Sometimes, however, it is an unintelligible mangle.

In some ways, agent-based models complement traditional analytic methods. Where analytic methods enable humans to characterize the equilibrium of a system, agent-based models allow the possibility of generating that equilibrium. This generative contribution may be the most mainstream of the potential benefits of agent-based modelling. Agent-based models can explain the emergence of higher order patterns - network structures of terrorist organizations and the internet, power law distributions in the sizes of traffic jams, wars, and stock market crashes, and social segregation that persists despite populations of tolerant people, etc.

2.4.2 Cellular Automaton

Cellular Automaton (CA), can be considered as a particular case of agent based model where the agents are fixed and contiguous surface elements. Ulam and Von Neumann (1961) state that a CA is a cellular entity that independently varies its states based on its previous state and that of its immediate neighbours according to specific rules. It is a spaced dynamic system where the variables (ex. land cover), time and space are discrete (Houet et al, 2006).

The state of a cell in the future time step can be defined mathematically as a function of the present state of that cell and its neighbours and a set of rules that express the dynamics or process between them.

A CA is defined by:

• Space: represented as an infinite and regular tessellation of cells of discrete states (generally a matrix).
• Set of states: is the set of possible values associated to the cells.
• Neighbourhood: corresponds to a set of adjacent cells.

• Transition rules: specified as a rule table that defines the next state of the cell for each possible neighbourhood configuration. They are uniformly applied to all cells at fixed time intervals.

CA models have been increasingly used to simulate land-use and land-cover changes due to their computational simplicity and their explicit representation of space and time. Typically, these models use the raster model, as defined in Geographic Information Systems, to represent geographic space (Moreno, et al 2007).

2.4.3 Elements of AB and CA Models

There are common elements and definitions for agent based and cellular automata models as has been described earlier, such as:

The Cell: defines the basic spatial unit of the system being modelled. The typical arrangement is a two dimensional grid, like a raster map, for urban modelling environments. But it is possible that the modelling environment is 1D like in models of traffic or 3D for some models of molecular gases interaction. The size of the cell is important since this represent the scale of the process that wants to be modelled. For urban environments the cell size can be 200*200 meters or less depending on the amount of information available to set up the conditions in the model. A common source of information for setting up such a model comes from remote sensors like satellites and aerial photographs to classify land uses and coverage. The selection of a smaller scale makes the system more complex and computational demanding. Although is possible to build a model at a house scale it must be kept in mind that urban expansion rarely as a result of an individual decision but as block of houses or housing plans from the municipalities. In the ABM technology the agents can access information regarding the environment condition in each cell.

The cell state: defines the attributes in the system. Each cell can take only one state at every time step. In urban modelling states can represent a property or urban land category such as residential, commercial, industrial etc. Every time the system is updated as a result of the analysis of the transition rules in the interactive neighbourhood, the state of each cell will be updated either change or stay in the same state. For Agent based models the state is equivalent to define the role that each agent is playing in the simulation.

The neighbourhood is the set of cells that are considered to play a role or influence the state of the cell that is analyzed. In 1D model the adjacent cells are considered in the analysis. In two dimensional space several configuration of neighbourhoods have been evaluated, the more typical set up is the Von Newman and Moore neighbourhood. The Von Newman neighbourhood includes 5 cells in the analysis; see figure 2.1 A, the Moore neighbourhood considers the 8 cells around the central cell of analysis in a 3*3 array. There is an extended Moore neighbourhood which consists of a 5*5 array. See figure 2.1 B and C.

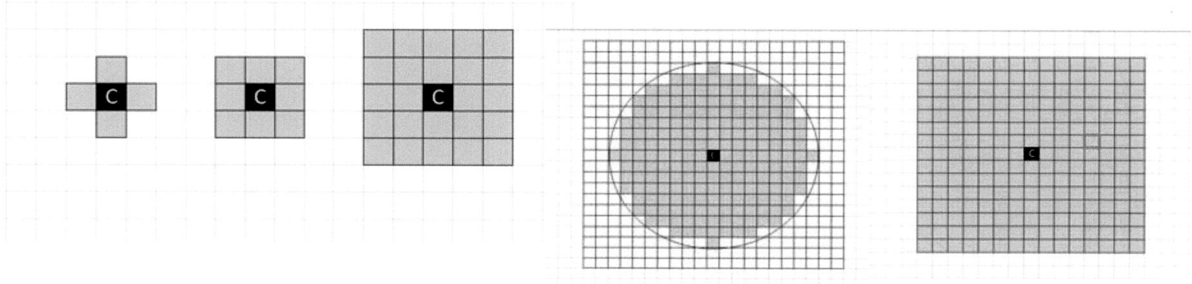

Figure 2.1. A. Von Newman Neighborhood, B. Moore 3x3 Neighborhood, C. Extended Moore 5x5 Neighborhood, D. Circular Neighborhood 8 Cells diameter, E. Rectangular Neighborhood proposed by Wu (1996).

In the ABM the neighbourhood is define according with the interaction of each agent with the other agents it encounters in his path, since the agents are dynamically moving in the cell space or environment.

The transition rules directly represent the process that is being modeled and as such are the key elements in CA and ABM. The rules define how each cell or each agent is going to change its state in response to the current conditions in the neighbourhood or surrounding agents. Thus the rules represent the dynamics of the system.

In cellular automata these rules normally are a set of IF-THEN statements, such as:
"If something happens in the neighbourhood of a cell, then a change will happen to that cell in the next step".

For example in the "Game of Life" the natural behavior of live and death can be modeled using a set of three rules:

If there are 2 or 3 live cells in the Moore neighbourhood of a live cell, then that cell stays alive in the next generation.

If there are less than 2 or more than 3 live cells in the Moore neighbourhood of a live cell, then that cell dies in the next generation.

If there are exactly 3 live cells in the Moore Neighbourhood of a death cell, then that death cell becomes alive in the next generation.

The time defines the temporal dimension were the CA system exist. From the definition of CA all the cells states are updated at every time step, which depend on the state of the cell itself and the state of the cells in the neighbourhood of analysis in the previous time step. Some models are configured by starting the model from known spatial data sets that is available, and then letting the model run for a number of iterations until the simulated results "fit" with another set of data at the ending time. These kind models were not configured temporally. There are other examples such as constrained cellular automata where there is an external model that constrained the number of cells per state in time (Engelen and White, 1996).

2.4.4 Agent based models for water management

Managing water systems is a complex task, having to cope with various water-related activities and conflicting user perspectives within a specified geographical area – basin, catchment, watershed etc. Typically, there are several stakeholders involved, and their different, typically contradictory goals must be seriously taken into consideration. Water supply for domestic uses, agricultural or industrial use, environmental issues and recreation or amenity provision are only some of the activities the different stakeholders may be involved in (Tzima et al, 2006).

Urbanization and overexploitation of the water resource often leads to water scarcity by quantity and/or quality, aggravating the situation to a socio-economic power struggle. In such complex and multidimensional cases, management strategies not only have to balance water demand and supply, but also find solutions that meet the approval of all users. Issues like the prioritization of users, the construction of water tariffs, the protection of ecological reserves, the compliance with the economic objectives and the legislative context needs to be equally considered.

This process of conflict resolution has to be done within a system of strongly coupled biophysical, social and economic entities, where the impact of certain strategies cannot be assessed by long-term studies or experimental manipulations alone, making it complex.

Simulation models, and particularly agent-based simulation models, are tools that can facilitate overcoming these limitations. Such tools can be used to evaluate the possible effects of different management plans. It is important to highlight that ABMS tools are not developed to forecast the exact state of the modeled system, but to explore how the system will evolve in view of a possible future. Table 2.1 presents a summary of agent based models used for water management.

Table 2.1. Agent based models for water management

Model	Description
SHADOC	SHADOC developed by Barretou et al, 2000, is a MAS seeking to examine the viability of, currently underutilized, irrigated systems in the Senegal River Valley. Based on the assumption that the interaction of the different system components has a large impact on its viability, the model focuses on rules used for credit assignment, water allocation and cropping season assessment, as well as on organization and coordination of farmers.
SINUSE	SINUSE, addresses the problem of integrated management of the Kairouan water table, located in Tunisia, which has been continuously decreasing for more than 20 years. It is an attempt to model the observed system dynamics and explore the effects of different kinds of intervention. Feuilliette et al, 2003.
CATCHSCAPE	Becu et al. 2003 have developed CATCHSCAPE, an agent-based model for the management of Mae Uam, a small catchment in Northern Thailand. The model intends to explore the impact of upstream irrigation management on downstream agricultural viability in an environment where biophysical and social factors are a source of conflict. Thus, in an attempt to foster the achievement of negotiated settlements to such conflicts, it simulates the whole catchment features as well as farmer's individual decisions.
AWARE	Action-research and Watershed Analyses for Resource and Economic sustainability (AWARE) is a simulation tool that models the dynamics of catchment level water management in South Africa, . The water management approach promoted by the National Water Act of South Africa relies on a licensing process, through which

Model	Description
	water use authorizations are allocated to various competing groups of users. In this context, AWARE is meant to be a support tool to evaluate alternative scenarios representing potential water management strategies.
MANGA	MANGA is a tool aiming to assist decision-makers in the difficult task of collective management of water resources it was developed by Le Bars et al, 2005. It provides a simulation environment for testing the consequences of various water allocation rules, in order to identify an acceptable compromise. Rule consequences depend on agricultural constraints, actors' behaviours, and confrontation of their decision rules with other actors.
The Bali Model	The Lansing-Kremer model (Lansing et al, 1993) simulates the irrigation system of the Oos- Petanu watershed in Bali. It involves the representation of the various water flows, the topology of rice terraces, as well as the coordination procedure used by local farmers for water allocation and control pests. The aim of the model is to prove that among the various levels of coordination for water sharing, the temple level, traditionally used, maximizes the production of rice.
The Lake Model	Eutrophication is a widespread and growing problem of hydro-systems that is caused by an excess input of nutrients, like phosphorus. The model presented in Janssen, 2001; specifically focuses on the management of lake eutrophication and explores the lake dynamics in relation to the behaviour of agents using phosphorus for agricultural purposes in the area.

2.4.5 Agent-Based Models for Water demand and supply management

Water management in urban areas is a challenging procedure during which, environmental, economic, social and political parameters have to be considered. Population growth, technological development, urbanization trends and climate change form a landscape of water scarcity, where water demand needs to be controlled and reshaped. Examples of water demand control mechanisms include variable water tariff schemes, exhortations, public campaigns and promotion of water-saving technologies.

The common objective of all these demand management strategies is water conservation, or in other words, the encouragement of customers to make more efficient use of the resource. Their evaluation process aims on calculating future water demand changes and must therefore take into account all the factors affecting water demand behaviour. In this context, social behaviour is a critical aspect, as most demand-side strategies focus directly - or indirectly - on human behaviour.

Water demand management is inevitably influenced by both environmental and socio-economic factors. Thus, the need for tools that can model the water management dynamics in an integrated way is evident. Unlike traditional statistical or econometric approaches, these tools should take into account social interaction and human behavior, and achieve the difficult task of promoting public participation.

In achieving this bilateral goal, agent-based modelling is a promising approach. The following sections briefly present some of the agent-based models for water management in urban areas found in the literature.

The Firma Thames Case

The "Thames" model was developed as part of the FIRMA project, and particularly the case study concerning the south of England, and aims at balancing supply and demand, in order to ensure effective water use during periods of climate change. The modelling procedure involved the development of two versions of the system, the second being more specific and realistic, based on the feedback received from stakeholders.

Modelling approach

The first version of the system, implements a hydrological model that determines the amount of water in the soil and a society of agents representing domestic users with various monthly water consumptions. A policy agent represents the policy authorities and determines restrictive exhortations when drought conditions prevail.

Each household agent determines its frequency of water consumption events and the quantity per event. To do so, agents employ a decision-making process based on endorsement schemes and chose among alternative behaviors according to evidence, social pressure and public authority exhortations. The latter, are recommendations by the policy agent, concerning the frequency of each activity as well as consumption per use event. These recommendations become stricter when drought conditions persist and cease when the soil water levels recover.

In the second stage of water demand modelling the random frequencies and volumes of water per consumption activity, were replaced by actual statistical data. The model was also extended to specify ownership of appliances related to specific water consumption activities, thus making it possible to investigate the effects of technological change.

The FIRMABAR Case

FIRMABAR is an agent-based simulator developed as part of a second case study of the FIRMA project, in the Metropolitan Region of Barcelona (MRB). It is an integrated tool for the evaluation of alternative supply and demand policies under different climatic and technological scenarios, taking into account the important changes in the territorial model of the region. Similar to the "Thames" model, its design and validation were based on participatory processes with a platform of representative stakeholders.

Modeling approach

The modeling of the system is based on the assumption that urban dynamics play a central role in water consumption behavior. Thus, space is represented explicitly as a grid and follows cellular automata rules. The agent model involves the following actors:

i) The families that incorporate social attitudes and lifestyles and compute their maximum expected water demand and real consumption.

ii) The real estate companies that build new housing around the municipalities and act as intermediates in the second-hand market. This modeling choice reflects the trend "driving the territorial model from the compact city to diffused patterns of urbanism".

iii) The municipalities that correspond to the different districts in the studied area - each of the latter having an initial urban development plan.

iv) Regional government that decides about water prices and infrastructure investments calculates the water stock and enforces various supply-and-demand policies in the area.

Families have to make decisions on house movement and water consumption. House movement depends on several parameters such as the size of the house versus size of the family, social class in the neighbourhood and the evolution of prices for new houses versus the second-hand market. Based on these factors, families may decide to migrate, thus producing a new territorial model.

The second decision a family has to make concerns its water consumption and partly depends on individual level rules and preferences (water price and gross income, maximum demanded water, housing type and appliance technology, social class and size of the family). Another factor affecting this decision is the social attitude towards emergency situations of drought or scarcity. Finally, water consumption is adapted to the local habits by a local mechanism of imitation.

2.4.6 Agent Based Models for Urban and Peri-Urban Water Management

Peri-urban areas around rapidly growing cities are, most of the times, characterized by a ''patch structure'' in terms of land use, that ranges from urban infrastructure to strictly rural and agricultural use. On the other hand, illegal settlements in these areas are a very common case and suffer from a lack of basic infrastructure and public facilities, thus constituting major non-point pollution flow sources. These dynamics directly affect the hydrological processes of the whole area, by changing the permeability of the soil, the runoff coefficients, the distribution of peak flows in the natural streams, the degradation of the water quality in drinking water reservoirs and aquifers. In many cases, this situation is already leading to water use restrictions and open conflicts.

The management of such complex and dynamic systems must take a more integrative approach and jointly consider all the hydrological and social processes involved. Agent-based models have been tested in this kind of participatory approach for ecosystem management and proved to be interesting and effective tools in the modeling and implementation phases. In the case of peri-urban areas, they can provide a way to structure and study the interactions between conflicting land uses and the competition for water availability among urban demand, agriculture irrigation, industry and recreational activities.

The Sao Paulo Model

The model reported by Ducrot et al, 2004 is MAS attempting to represent the relationships between urbanization dynamics and land and water management in peri-urban catchment areas. It was inspired from the spring areas of Sao Paulo city, which is the main drinking water reservoir of a great metropolitan area, suffering from high urban pressure and problems of pollution connected to land use and rain. The model takes a combinatorial approach, using cellular automata, spatial passive entities and communicating agents.

Modelling approach

The model's architecture involves a Cellular Automata (CA) layer to spatially represent the hydrological model and its dynamics. The agents are used to represent the behaviors of stakeholders and the decision-making processes, affecting water management in the area. Land-use changes and urbanization are specifically taken into account.

In order for the model to be spatially explicit, the catchment area is represented as a grid, with each cell having a unique value of land cover: reservoir, river, favelas, residential building, industry, tourism infrastructure, irrigated crop (horticulture), or non-irrigated crop (cereals). Cells are also aggregated into three municipalities and one reservoir, the latter having a specific water level and pollution rate.

There are two types of land-user agents, farmers (or producers) and urban owners (or speculators), all initially attributed with a plot, a cashbox and family needs.

Farmers all grow crops in their plots, choosing among horticultural crops that need to be irrigated, cereal crops (not irrigated), and fallow. In the other hand, urban owners employ one of the following strategies: i) using the plot for recreational purposes, ii) using the plot for speculation purposes, waiting for a higher price, or iii) developing a profitable activity on their plot, such as tourism or industry.

Any plot may be left in "fallow" state and changes in land use are possible, depending on owner's parameters and on the local configuration around a plot. Land-use dynamics are central in the model and rely on two different driving forces: transition rules between cells, and agents' decision-making. Examples of changing land-use include: (i) unoccupied cells in the immediate surroundings of a favela becoming urbanized, (ii) the rural land-use model changing according to the farmers' cropping choices and (iii) investments changing plots into touristic or industrial settlements.

The model's migratory dynamics are summarized in two processes. An immigration process involves new land-user agents being added to either the farmers or the speculators population. On the other hand, land-users owning no plots are removed from the model, in a process representing emigration.

Finally, a land market is organized every year, where indebted farmers have to sell their plots. The model adopts "priority sale to neighboring farmers to account for social links between farmers as well as for a preference to avoid land dispersal".

WULUM

Water Use and Land Use Model (WULUM) (Zellner, 2007) is an agent-based model investigating the relationship between land use, water use and groundwater dynamics in the Monroe County in Michigan. The main objective is to link these processes in an integrated model and evaluate their effect on groundwater levels.

Modelling approach

The physical model in WULUM is spatially explicit. A grid is used, where each cell contains information including groundwater, forest, soil quality, roads, zoning restrictions and municipal water coverage. Cellular automata transitions rules are used to create the regional west-east groundwater gradients with points of recharge and discharge. Under the same principle, farm cells may become populated by residents or quarries, depending on residential preferences for location, existing development and zoning.

The agent model involves three water-extracting actors: residents, stone quarries and farmers. Agents are located on the grid and perform several water extracting activities, thus changing the landscape and the levels of groundwater.

These changing conditions then provide feedback to the decisions space of the agents. As far as the water resource dynamics are concerned, the water aquifer's level depends on the natural renewal rates and the degree of development. Moreover, the climatic (precipitation), hydrological (groundwater flow) and demographic data (annual rates of residential growth) used in the model are based on literature and expert knowledge about the area.

Tijuana's Border Town Model

This model developed by De Leon , 2007, simulates various socio-economic realities of low-income residents of the City of Tijuana for the purpose of creating propositional design interventions.

The Tijuana Bordertowns model allows input of migration rates and border crossing rates relative to employment and service centers in order to define population types (i.e. migrants or full-time residents). Population densities relative to block sizes are adjustable via the interface in order to steer the simulation toward specific land-uses such as urban centers or peripheral (rural) development. The rate of residential building is adjustable based upon relative community size, land value (approximate), required (per-capita) capital and the carrying capacity of (potentially) existing infrastructure.

Modelling Approach

A CITYSCAPE is generated, spreading out from a city center. Each patch is assigned a land-value, and a level of electrical, water and transportation service. A road network is drawn, and maquiladoras or industrial areas are placed at the edge of the city.

An initial set of migrants are created at the edge of the city on "irregular" patches, meaning those patches with a low land-value, near water and away from maquiladoras. A second set of migrants is created in the neighboring patches. This establishes the base population of the irregular settlements. The migrants also keep their citizenship, such as Jaliso or Oaxaca.

Then each migrant is assigned a living-expanses values, which is determined by the value of the land they occupy, food and other utility costs, as well as the electrical, water, and transportation. Food and other utility costs are constant for all agents. The electrical and water costs are determined by the patches values. Transportation is determined by their distance to service centers (like shopping centers), the distance to the maquiladoras they work at, and the access to transportation services.

With each model tick, new migrants move into patches next to existing migrants, some migrants cross the border, some migrants move into nicer locations once they have sufficient savings, and all of them participate in the economy, earning and spending money, as well as saving if possible.

New migrants will enter in unoccupied spaces adjacent to migrants who came from their home state.
Migrants that move will look for a patch in their area with electrical and water services, which is affordable to them.

Groups of migrants form colonias. The size of these colonias is determined by the COLONIA-SIZE slider. The larger the value, the larger the colonia. Colonias with sufficient density will be targeted for regularization. New electrical, water and transportation services will be developed for them.

This model explores the migration of population and the development and growth of the economy in the city. This is a preliminary version of the model and there are some capabilities that are not yet fully developed or implemented, such as the extension of the city infrastructure for the regularization of the colonias with high potential.

The Cities Model

The Cities model has been developed in the framework of the project "Procedural modeling of cities" at the Center for Connected Learning (CCL) and Computer Based Modeling, Lechner et al , 2004. This model allows a user to create a terrain and environment in which a set of builders will create a city. Users can interrupt the city build and change the environment and then continue the city creation simulation. The city consists of roads, parcels (sets of patches which are developed as a single unit), and buildings. Parcels are zoned for specific types of usage, commercial, industrial, residential, or park.

Modelling Approach
The user paints parameters onto the terrain, such as elevation, road grid constraints, and "honey" which are hot spots that attracts specific type of developers. The user may also create multiple city seeds, or move the existing one. The user may also draw primary roads and link them to the edge of the terrain, to simulate the city's major routes.

In addition the user can also set some of these parameters at a global level, affecting all patches. There are also global variables constraining the number of developers, the ratio of land-use between the development types, minimum parcel sizes for different land uses, and many other factors.

There are agents sets with role of builders for each type of development, such as residential, commercial, industrial, and park. These agents move through the terrain, grouping patches into parcels and then attempting to "develop" them by putting a new building upon the parcel, or increasing the size of the current building. Road builders move through the terrain building roads between areas, thus increasing their value.
At any time, the user can stop the model, paint new parameters onto the terrain, free up terrain that has been developed, draw primary roads, or change the land use ratios. When the model is restarted, builders will respond to these changes.

In this model there is a set of agents that has the role to search the modeling space for patches that are not connected or far away from the road network, thus extending and adding new roads to the model according with the parameters that affect the accessibility and that are user specified.

This model presents a good example of an agent base system to generate urban dynamics and complexity. It is a tool the gain inside in the process that drives urban development rather than a prediction tool. Nevertheless the authors reported that this model can generate land use distribution that realistically represents the urban landscape of cities like Chicago in the USA. There is not report of the model being applied to a real case dataset.

The SLEUTH model
The SLEUTH model is a cellular automaton model, developed with predefined growth rules applied spatially to gridded maps of the cities in a set of nested loops, and was

designed to be both scalable and universally applicable. Urban expansion is modeled in a modified two-dimensional regular grid. Maps of topographic slope, land use, exclusions, urban extents, road transportation, and a graphic hill shade layer form the model input. The model was first developed and applied in the San Francisco Bay area in the United States (Clarke, Hoppen, and Gaydos 1997). The name of the model came from the six input data layers, namely Slope, Land cover, Exclusion, Urbanization, Transportation, and Hillshade.

The model applies four types of urban land-use change: spontaneous growth, new spreading centre growth, edge growth, and road-influenced growth. A spontaneous growth occurs when a randomly chosen cell falls close enough to an urbanized cell, simulating the influence of urban areas on their surrounding land; a new spreading centre growth spreads outward from existing urban centres, representing the tendency of cities to expand; an edge growth urbanizes cells that are flat enough to be desirable locations for development even if they do not lie near an already established urban area; and a road-influenced growth encourages urbanized cells to develop along the road network.

The four types of urban growth are applied sequentially during each growth cycle and are controlled through the interactions of five growth coefficients: diffusion, breed, spread, road gravity, and slope (Clarke, Hoppen, and Gaydos 1997; Clarke and Gaydos 1998). The first four coefficients describe the growth pressure in the urban system. For instance, the diffusion coefficient determines the overall outward dispersive nature of the distribution; the breed coefficient specifies how likely a newly generated detached settlement is to begin its own growth cycle; the spread coefficient controls how much diffusion expansion occurs from existing settlements; and the road-gravity factor denotes the attraction of new settlements toward and along roads.

Resistance to growth is incorporated through the slope coefficient, which captures the effect of steep slopes on restricting development. In addition, resistance is also applied through an excluded data layer that identifies areas that are wholly (e.g.,water or parks) or partially (e.g., restrictive zoning) excluded from development. All five coefficients are calibrated to control the growth rate so that growth will not become unusually high or low. The overall rate of urban growth is the sum of the four types of growth.

The SLEUTH model is implemented in two general phases: a calibration phase, in which the model is trained to replicate historic development trends and patterns, and a prediction phase, in which historic trends are projected into the future. The model can be used to simulate the non-urban to urban conversion; it can also model the process of multiple land-use change.

This model has been applied and calibrated in real cases, for example Silva and Clarke, 2002 developed a SLEUTH model for Lisbon and Porto in Portugal. The model was successfully calibrated and validated, capturing the dynamics of the cities and was proposed to be used for planning and evaluation of future scenarios of development.

Moland Model

The aim of MoLAND is to provide a spatial planning tool that can be used for assessing, monitoring and modeling the development of urban and regional environments. Moland was a project that was initiated in 1998 (under the name of MURBANDY - Monitoring Urban Dynamics) with the objective to monitor the developments of urban areas and identify trends at the European scale. The work includes the computation of indicators and the assessment of the impact of anthropogenic stress factors (with a focus on expanding settlements, transport and tourism) in and around urban areas, and along development corridors. JRC, 2009 (accessed via web)

Since 2004, MoLAND is contributing to the evaluation and analysis of impact of extreme weather events, in the frame of research on adaptation strategies to cope with climate change. The MoLAND methodology has been applied to an extensive network of cities and regions for an approximate total coverage in Europe of 70,000 km2.

The MOLAND urban growth modeling tool was developed by the company RIKS (Research Institute for Knowledge Systems, Ltd) The model is part of RIKS's Metronamica modeling framework, which is based on dynamic spatial systems called cellular automata. www.riks.nl. The model takes as input different types of spatially referenced digital data:

Land use maps, showing the distribution of land use types in the area of interest.
Suitability maps, showing the inherent suitability of the area of interest for different land use types. These maps are created using an overlay analysis of maps of various physical, environmental and institutional factors.

Zoning maps, showing the zoning status (i.e. legal constraints) for various land uses of the area of interest. These maps are dervived from existing planning maps (e.g. master plans, zoning plans, designated areas, protected areas, historic sites, natural reserves, land ownership).
Accessibility maps, showing the accessibility to transportation networks for the area of interest, and they are based on the importance of access to transport networks for the various land uses.

Socio-economic data for the main administrative regions of the area of interest, comprising demographic statistics (i.e. population and income), and data on production and employment for the four main economic sectors (i.e. agriculture, industry, commerce, and services).

The main components of the model are:

- A two-dimensional grid or cell space, each grid-cell having its own unique set of attributes (i.e. land use, suitability, zoning, accessibility, socio-economic);
- A cell neighborhood, consisting of a circular area of radius eight pixels around each cell;
- A set of discrete cell states (i.e. 24 MOLAND land use classes or the Corine land use classification);
- Transition rules (describe the effect of neighboring cells on central cell).

The socio-economic information is used to build a regional model that accounts for the demand of the different land uses in the future or modeled time frame. Thus, the regional model constrains the local model or cellular automata by assigning a number of cells according with the demand. Once all the cells for a particular land use are assigned no other cell can change to that land use even if it has the highest potential, instead it will get the immediate land use class in the highest potential list.

Moland has been applied in several European cities to study the urban dynamics and patterns, for spatial planning and hazard mitigation, strategic environmental assessment, the application of sectoral policies and their spatial impact. Figure 2.2 shows the application of Moland in Dublin urban area. Barredo et al, 2003, Engelen et al 2007, RIKS, 2009.

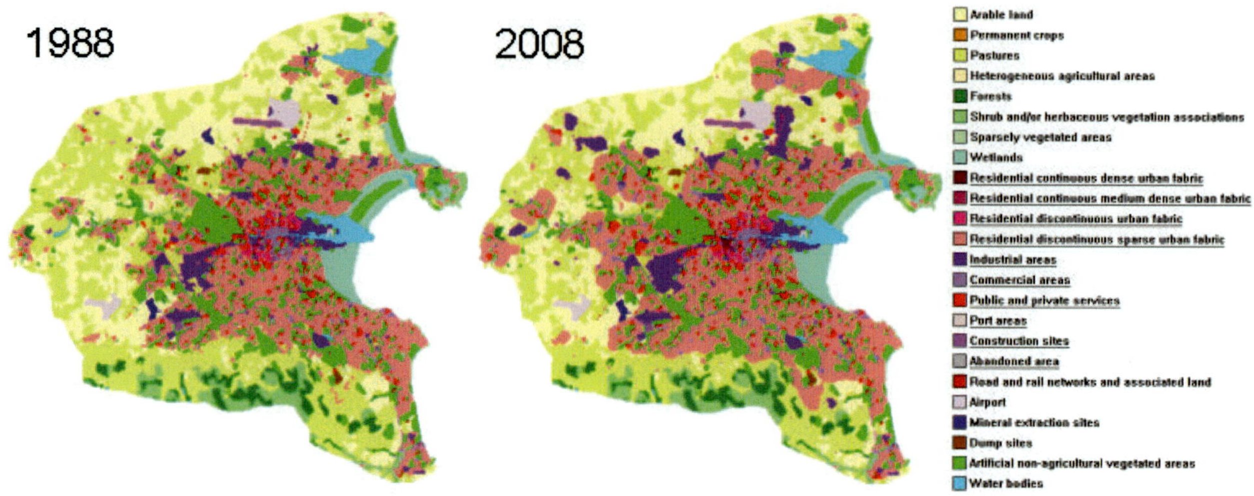

Figure 2.2. Urban growth in Dublin present and predicted land use change.Source: Engelen et al 2007

The output from the MOLAND urban model are maps showing the predicted evolution of land use in the area of interest, over the time. By varying the inputs into the MOLAND urban model (e.g. zoning status, transport networks), the model can be used as a powerful planning tool to explore in a realistic way the future urban and regional development of the area of interest.

The MedAction PSS model

The MedAction system was developed as part of an EU project "Policies to combat desertification in Northern Mediterranean Region". It is a dynamic spatial integrated model, which integrates 15 individual models with different modeling paradigms and temporal resolutions varying from minutes (for the rainfall and erosion models) to a year (for the land use and crop choice models). It has a finest spatial resolution of 1 ha grid cells and can therefore incorporate detailed spatial characteristics. The model is implemented in the GEONAMICA framework developed by RIKS in the Netherlands, similarly of the Moland model describe in the previous section.

The MedAction PSS reuses to a large extent models incorporated in its predecessor, MODULUS. The latter, in turn, is based on past research carried out in a number of EC-funded projects. The models that were included in the Core of MedAction are: Climate and Weather, hydrology, sedimentation, salinisation, water demand and usage, water resources, land use, profit and crop choice, dynamic suitability, plant growth, natural vegetation, land management. Van Delden et al, 2007 and Kok et al, 2007.

The integration of the processes and models provides a tool to uncover the behaviour of the system analyzed as the result of its autonomous dynamics largely determined by the human agents active in the system, subsidies and other policy measures imposed on the agents, and the exogenous drivers, climate change, technological change, demographic growth, and market forces. While the autonomous dynamics are very much core elements of the processes represented, policy measures and exogenous drivers impact on one or more model in the system. Impacts are passed on from one to the other models as a result of the many feedback loops. A high degree of integration was achieved in this model by providing more feedback loops between the models and by incorporating the dynamically suitability model that directly affects the change of land use patterns.

This model is in a demonstration version and the authors indicate the there is further work to be done to validate the model and released a version as an end user application.

2.5 Conclusion

This chapter presented the basic concepts and theoretical background related to integrated urban water modelling. It also presents the basic concepts of cellular automata and agent base models. A description of several experiences reported in the literature regarding the application of agent base modelling to urban growth and water management is given.

The revision of the experiences show that this modelling paradigm is rather new and the application have been developed in the last 15 years. Nevertheless, the experiences show that agent based models can replicate the complexity and dynamics of urban growth in real cases. The different experiences also shows that there has been advances in the technique and approach, there are several authors testing different neighbourhood sizes and shapes and evaluating different techniques to formulate the transition rules.

The amount of land use classes considered in the models is limited to main drivers or stakeholders. The classes at the urban level can be considered as urban or non urban activities. This is understandable since we know very little about the main forces and drivers that shape our modern cities.

Almost all the experiences considered land use classes or categories that are assigned to every grid in the model. But there is little information about population dynamics and building classification at the urban level, these characteristics are important for urban water management since water use and patterns are associated with building types, population density, population income, etc.

The selection of the neighbourhood is pragmatic. There is very little information or recommendations from the theoretical or experienced point of view, in the experiences reviewed. We know that the size of the neighbourhood will affect the shape of the clusters in the final simulation. And that the size depends very much of the processes involved in the phenomena being modelled. For urban scale the approach used by Riks has shown good results in practice.

Most of the models are not set temporally. This means that the simulation is run for a consecutive number of iteration until the results match a dataset that is used for calibration. The only model that considers time is Moland or the Geonamica framework developed by RIKS. Their approach used socio-economic information to set-up a regional or national model that imposes constraint in the growth per land use, according with the plans and expectations at a macro level.

The integration of urban growth models with urban water management has been explored in some the experiences presented here. The experiences report that the integration is hard to achieve and the models become even more complex. Most of the integration is done by incorporating the outputs of numerical models as inputs in the agent based models in the form of suitability maps with certain weight in the decision making process. The MedAction experience report a full integration with several physically based models, the integration was achieved by developing a dynamically suitability model that handles the changes caused by certain actions in the output of the individual models, then creating more feedback loops to other models that can be affected and finally building a new set of suitability map per time step in the simulation.

3 Framework to Model Cities Future Growth

3.1 Introduction

Land-use change is a complex, dynamic process that links together natural and human systems. Since this system act as an intermediary between the atmosphere and the water systems, understanding its dynamics is important for several environmental problems (Meyer and Turner, 1994). The big changes conducted in the tropical forest to transform the land into agricultural uses have a direct impact on biodiversity, lost of soil and nutrients, it is possible that this is already having an impact in a global scale and may be linked to climate change (Lambin et al., 2003).

Land-use change is also one of the important factors in the climate change cycle and the relationship between the two is interdependent; changes in land use may affect the climate whilst climatic change will also influence future land-use (Dale, 1997; Watson et al., 2000).

Modelling land-use change helps understand the processes of urbanization and it is important as a source of information for politicians and planners to assess future conditions under different scenarios of urban growth. Land-use change models are tools to understand the causes and consequences of land-use change (Verburg et al., 2004). Recent inventories of operational models for land-use change are numerous. Briassoulis (2000) offers a extensive discussion of the most common land-use change models and their theoretical background. One of the most important distinctions refers to static as opposed to dynamic models. Static (or cross-sectional) models directly calculate the situation at a given point in time, whereas dynamic models work with intermediate time-steps, the latter, therefore, takes possible developments during the simulation period into account, providing a richer behaviour and the possibility to better mimic actual spatial developments and the incorporation of several variables that may influence these dynamics (Koomen et al. 2007).

3.2 Modelling of Land use change

DINAMICA EGO is used as a simulation platform for the urban dynamics model, Soares et al, 2011. DINAMICA employs, as input, a set of maps, including the initial and final map of land use, also known as landscape maps, where a landscape is viewed as a bi-dimensional array of land use types, and two sets of ancillary maps: the static and dynamic variables, the latter so named because they are updated by the model iteration. The sets of variables control the location of changes. These variables are combined by summing their Weights of Evidence (Goodacre et al. 1993; Bonham-Carter, 1994 and Soares-Filho et al. 2011), to produce a transition probability map, which depicts the most favourable areas for change (Soares-Filho et al. 2002, 2004 and 2011).

3.2.1 The method of Weights of Evidence

The Weights of is a bayesian method, in which the effect of the spatial variable on a transition is calculated independently of a combined solution. The weights of evidence represents each variables influence on the spatial probability of transition i to j. and are calculated as follow:

The probability to find a residential class (R) given the presence of water supply service (S), evidence, is determine by the following equation:

$$P\{R/S\} = \frac{P\{R \cap S\}}{P\{S\}}$$

Where:

P{R/S}: Conditional probability that the transition event R occurs, given the presence of the variable or evidence S.

P{R ∩ S}: corresponds to the intercepted area between R and S.

P{S}: Area covered by the water supply service.

This can also be expressed by considering the total amount of cells in the area of interest, N. The previous equation is as follows:

$$P\{R/S\} = \frac{N\{R \cap S\}}{N\{S\}}$$

To obtain a relation for the probability 'a posteriori' of occurrence of an event R in terms of the known probability 'a priory' and a multiplication factor, it is possible to say that the conditional probability of the event R to be present within the area of evidence S is given for:

$$P\{S/R\} = \frac{P\{S \cap R\}}{P\{R\}}$$

Because P{S ∩ R} is the same as P{R ∩ S} it is possible to combine the equations to determine P{R/S}, as follows:

$$P\{R/S\} = \frac{P\{R\}.P\{S/R\}}{P\{S\}}$$

A similar expression can be derived for the probability 'a posteriory' of occurrence of the event R given the absence of event S, $\bar{S}$, is given by:

$$P\{R/\bar{S}\} = \frac{P\{R\}.P\{\bar{S}/R\}}{P\{\bar{S}\}}$$

The same equations presented here above can be rewritten in the odds form. Odds are defined as the fraction of the probability that an event will occur and the probability that it will not occur. The weights of evidence method use the natural logarithm of the odds, also known as log odds or logits. To obtain the equation in its odds form both sides will be divide by $P\{R/\bar{S}\}$, as follows:

$$\frac{P\{R/S\}}{P\{\bar{R}/S\}} = \frac{P\{R\}.P\{S/R\}}{P\{\bar{R}/S\}.P\{S\}}$$

According with the definition of conditions probability:

$$P\{\bar{R}/S\} = \frac{P\{\bar{R} \cap S\}}{P\{S\}} = \frac{P\{S/\bar{R}\}.P\{\bar{R}\}}{P\{S\}}$$

By substituting, this equation in the previous, it is obtained:

$$\frac{P\{R/S\}}{P\{\bar{R}/S\}} = \frac{P\{R\}}{P\{\bar{R}\}} \cdot \frac{P\{S\}}{P\{S\}} \cdot \frac{P\{S/R\}}{P\{S/\bar{R}\}}$$

In the odds form is:

$$O\{R/S\} = O\{R\} \cdot \frac{P\{S/R\}}{P\{S/\bar{R}\}}$$

Where

$O\{R/S\}$: conditional odds (a posteriory) of R given S

$O\{R\}$: odds a priory of R, and

$P\{S/R\}$, $P\{S/\bar{R}\}$: are known as the sufficiency ratio (LS).

By taken the natural logarithm at both sides:

$$\text{logit}\,\{R/S\} = \text{logit}\,\{R\} + W^{+}$$

Where W^{+} is the positive weight of evidence.

A similar algebraic formulation can be done to obtain the odds expression for the conditional probability of event R given the absence of evidence S. this is given by the following expression:

$$O\{R/\bar{S}\} = O\{R\} \cdot \frac{P\{\bar{S}/R\}}{P\{\bar{S}/\bar{R}\}}$$

$P\{\bar{S}/R\}$, $P\{\bar{S}/\bar{R}\}$ is known as necessity ratio (LN). By taking natural logarithms in both sides the negative weight of evidence if loge LN.

$$\text{logit}\,\{R/\bar{S}\} = \text{logit}\,\{R\} + W^{-}$$

LS and LN are also called likelihoods ratios. When the event and the evidence are positively correlated the value of LS is greater than 1. Meanwhile, the value of LN is within the interval [0,1]. If the evidence is negatively correlated with the event, LN is greater than 1 and the value of LS will be in the interval [0,1]. If there is no correlation between the evidence and the event, LS=LN=1, which means that the probability 'a posteriory' is equal to the 'a priory', the probability of the event is not affected by the presence neither the absence of the given evidence.

When the evidence from several maps is combined, the weights are computed for every map independently or combined in one equation. the conditional probability that an event occur, given the presence of two evidences S1(water supply service) and S2 (drainage service) is given by:

$$P\{R/S_1 \cap S_2\} = \frac{P\{R \cap S_1 \cap S_2\}}{P\{S_1 \cap S_2\}}$$

The above equation can also be written as:

$$P\{R/S_1 \cap S_2\} = \frac{P\{S_1 \cap S_2/R\}.P\{R\}}{P\{S_1 \cap S_2\}}$$
$$= \frac{P\{S_1 \cap S_2/R\}.P\{R\}}{P\{S_1 \cap S_2/R\}.P\{R\} + P\{S_1 \cap S_2/\bar{R}\}.P\{\bar{R}\}}$$

This is expression represent the Bayes theorem. According with this, two mutually exclusive hypothesis R and $\bar{R}$, with $P\{R\} + P\{\bar{R}\} = 1$. The effects of the interaction between S1 and S2 can be ignored if they are conditional independent among them. Therefore, a simplification can be done, because it allows to assess the effects of every map of evidence individually or with a combination by the multiplication (or sum in the log-linear case) of the factors from different maps together.

The conditional independency supposition can be expressed as follows:

$P\{S_1 \cap S_2/R\} = P\{S_1/R\}.P\{S_2/R\};$

Considering this simplification the equation for two maps can be written as follow:

$$P\{R/S_1 \cap S_2\} = P\{R\}.\frac{P\{S_1/R\}}{P\{S_1\}}.\frac{P\{S_2/R\}}{P\{S_2\}}$$

In Odds form is:

$$O\{R/S_1 \cap S_2\} = O\{R\}.LS_1.LS_2$$

Or by using the log-linear form of weight of evidence as follows:

$$\text{Logit}\{R/S_1 \cap S_2\} = \text{logit}\{R\} + W_1^+ + W_2^+$$

There are 4 possible ways to combine two evidence maps, the first one when both evidences are present, when the first evidence is absence and the second evidence is present , when the second evidence is absence and the first evidence is present, and when both evidences are absence. the combination can be written in the log-linear forms as follows:

$$\text{Logit}\{R/S_1 \cap \overline{S_2}\} = \text{logit}\{R\} + W_1^+ + W_2^-$$
$$\text{Logit}\{R/\overline{S_1} \cap S_2\} = \text{logit}\{R\} + W_1^- + W_2^+$$
$$\text{Logit}\{R/\overline{S_1} \cap \overline{S_2}\} = \text{logit}\{R\} + W_1^- + W_2^-$$

If the number of evidences is 3, then, the possible number of combinations are $2^3 = 8$. In general, for a given set of maps n, the number of combinations are 2^n. In the log-linear form it can be written as:

$$\text{Logit}\{R/S_1 \cap S_2 \cap S_3 \cap \dots S_n\} = \text{logit}\{R\} + \sum_{i=1}^{n} W_i^+$$

According to Bonham-Carter, 1994 some of the advantages of the Bayesian model are:

- Objectivity, which discourage the subjective selection of weighting factors.
- The possibility to combine multiples maps of evidence in a model that is easy to implement in a computational environment.
- The inclusion in the model of input maps with incomplete data.
- The possibility to use maps with multiple classes, where every class is treated as presence evidence.
- The model of uncertainties caused by variations in the weights and incomplete data.

Some disadvantages of the method can be:

- The combination of input maps assumes that they are conditional independent among them. A test to check the conditional independence must be run.
- The limitation to apply the model in cases when the event is well known.

Since weights of evidence only applies to categorical data, it is necessary to categorize continues gray-tone maps (quantitative data, such as distance maps, altitude, and slope). A key issue to any categorization process concerns the preservation of the data structure, DINAMICA uses a method adapted from Agterberg and Bonham-Carter (1990) to categorize continues maps.

3.2.2 Selection of Variables

The only assumption for the Weights of Evidence methods is that the input maps have to be spatially independent. A set of measures can be used to assess this assumption, such as the Cramer test and the Joint-Uncertainty Information (Bonham-Carter 2004). Correlated variables must be disregarded or combined into a third that will be used in the model. For this purpose Dinamica EGO contains a model to perform this statistical test between pairs of variables. Soares et al 2011. This model performs pair-wise test for the categorical maps in order to test the independence assumption. The indicators that are computed are Ch^2, Crammers, the Contingency, the Entropy and the Unicertainty joint information (Bonham -Carter, 1994). Figure 3.1 shows a log result of this model.

Transition: 2->4

First Variable	Second Variable	Crammer: Chi^2	Crammer: Crammer*	Crammer: Contingency	Entropy: Joint Entropy	Entropy: Joint Information* Uncertainty
MapStaticM3/D_Canals	MapStaticM3/D_MRivers	14728.8	0.0786558	0.291411	4.49735	0.0202522
MapStaticM3/D_Canals	MapStaticM3/D_MSewers	14898.6	0.0989092	0.298517	3.71437	0.0269561
MapStaticM3/D_Canals	MapStaticM3/D_MTrunks	2518.11	0.0572331	0.126942	3.64842	0.00439232
MapStaticM3/D_Canals	MapStaticM3/D_MotorWays	2726.24	0.0338887	0.130134	4.62421	0.00380734
MapStaticM3/D_Canals	MapStaticM3/D_PrimaryRoads	13010.6	0.0831766	0.276868	4.24584	0.0193111
MapStaticM3/D_Canals	MapStaticM3/D_Rails	13698.1	0.076296	0.28338	4.63206	0.0178198
MapStaticM3/D_Canals	MapStaticM3/D_SecundaryRoads	7929.32	0.0714122	0.220278	4.4242	0.0113578
MapStaticM3/D_Canals	MapStaticM3/Elevation	7047.8	0.0936219	0.204903	3.70547	0.0121361
MapStaticM3/D_Canals	MapStaticM3/slope	482.367	0.0344979	0.0596457	2.72283	0.00132564
MapStaticM3/D_MRivers	MapStaticM3/D_MSewers	12284.6	0.0892085	0.271506	3.61125	0.024948
MapStaticM3/D_MRivers	MapStaticM3/D_MTrunks	1505.62	0.0439603	0.0978268	3.53122	0.00268066
MapStaticM3/D_MRivers	MapStaticM3/D_MotorWays	5265.07	0.0439254	0.17821	4.48875	0.00690304
MapStaticM3/D_MRivers	MapStaticM3/D_PrimaryRoads	6647.9	0.0590578	0.200431	4.15305	0.0102539
MapStaticM3/D_MRivers	MapStaticM3/D_Rails	9547.1	0.0632798	0.238037	4.52218	0.0128006
MapStaticM3/D_MRivers	MapStaticM3/D_SecundaryRoads	9180.76	0.0763226	0.234616	4.29898	0.0133081
MapStaticM3/D_MRivers	MapStaticM3/Elevation	22116.4	0.164695	0.345581	3.53126	0.0393684
MapStaticM3/D_MRivers	MapStaticM3/slope	701.796	0.04133	0.071403	2.61254	0.00200965
MapStaticM3/D_MSewers	MapStaticM3/D_MTrunks	10112.2	0.116213	0.251507	2.75096	0.0236758
MapStaticM3/D_MSewers	MapStaticM3/D_MotorWays	23262.9	0.123025	0.362568	3.67722	0.0414255
MapStaticM3/D_MSewers	MapStaticM3/D_PrimaryRoads	20594	0.116161	0.344806	3.30967	0.0402945
MapStaticM3/D_MSewers	MapStaticM3/D_Rails	16497.5	0.104027	0.312489	3.71283	0.0313426
MapStaticM3/D_MSewers	MapStaticM3/D_SecundaryRoads	[illegible]	[illegible]	[illegible]	[illegible]	[illegible]

Figure 3.1 Test of correlation between pairs of maps.

Although there is not agreement on what threshold should be use to exclude a variable, if all the test highlight a high correlation between any pair of maps then it means that one of the maps is redundant and therefore not needed in the simulation. Soares et al, 2011. Almeida, 2003 based on Bonham and Carter, 1994 reports that values below 0.5 for the Crammer and the Joint information uncertainty indicators suggest less association than more. Despite the calculation of this statistical test of independence, Almeida, 2003 suggest that determining if a variable is independent against other variable is still arbitrary and that there is a lack of different cases where this methods has been applied in the literature to be used as a reference.

As a result, the spatial relationships calculated using the Weights of Evidence are used to parameterize and calibrate the simulation model with respect to the spatial configuration of changes.

3.2.3 Dynamics of land use transition

Another component of the model, the transition function, operates on the probability maps, and is constrained by the quantity of changes specified as input for each transition. This function depends on the higher probability cells, after having ranked them in a vector file. The quantities of changes are determined a priori through the calculation of a historical transition matrix. The transition matrix describes a system that changes over discrete time increments, in which the value of any variable in a given time period is the sum of fixed percentages of the value of the variables in the previous time period.

DINAMICA uses as a local CA rule, a transition engine composed of two complementary transition functions, the Expander and the Patcher (Soares-Filho et al. 2002). The software splits the cell selection mechanism into these two processes. The first is dedicated to the expansion or contraction of previous patches or clusters of a certain class, and is called the Expander. The second is designed to generate or form new patches through a seeding mechanism, and it is called the Patcher. For each transition, the percentage of transitions executed by the Expander function in relation to the Patcher function must be defined.

$$Transition(ij) = X * Expander + Y * Patcher$$

Where $X + Y = 1$.

According with Soares et al, 2002 in case that the number of transitions for certain cell state is not executed by the expander function after a fixed number of iterations, the remaining transitions are transferred to the patcher, so that, the total number of transitions reach the expected total value. This is a feedback mechanism between these two functions.

The size of new patches and expansion fringes are set according to a lognormal probability distribution. Therefore, it is necessary to specify the parameters of this distribution represented by the mean and variance of the patch sizes to be formed.

The Patch Isometry is a number varying from 0 to 2. The patches assume a more isometric form as this number increases. Figure 3.2 shows the required steps to set-up the cellular automata model of land use change within Dinamica Ego.

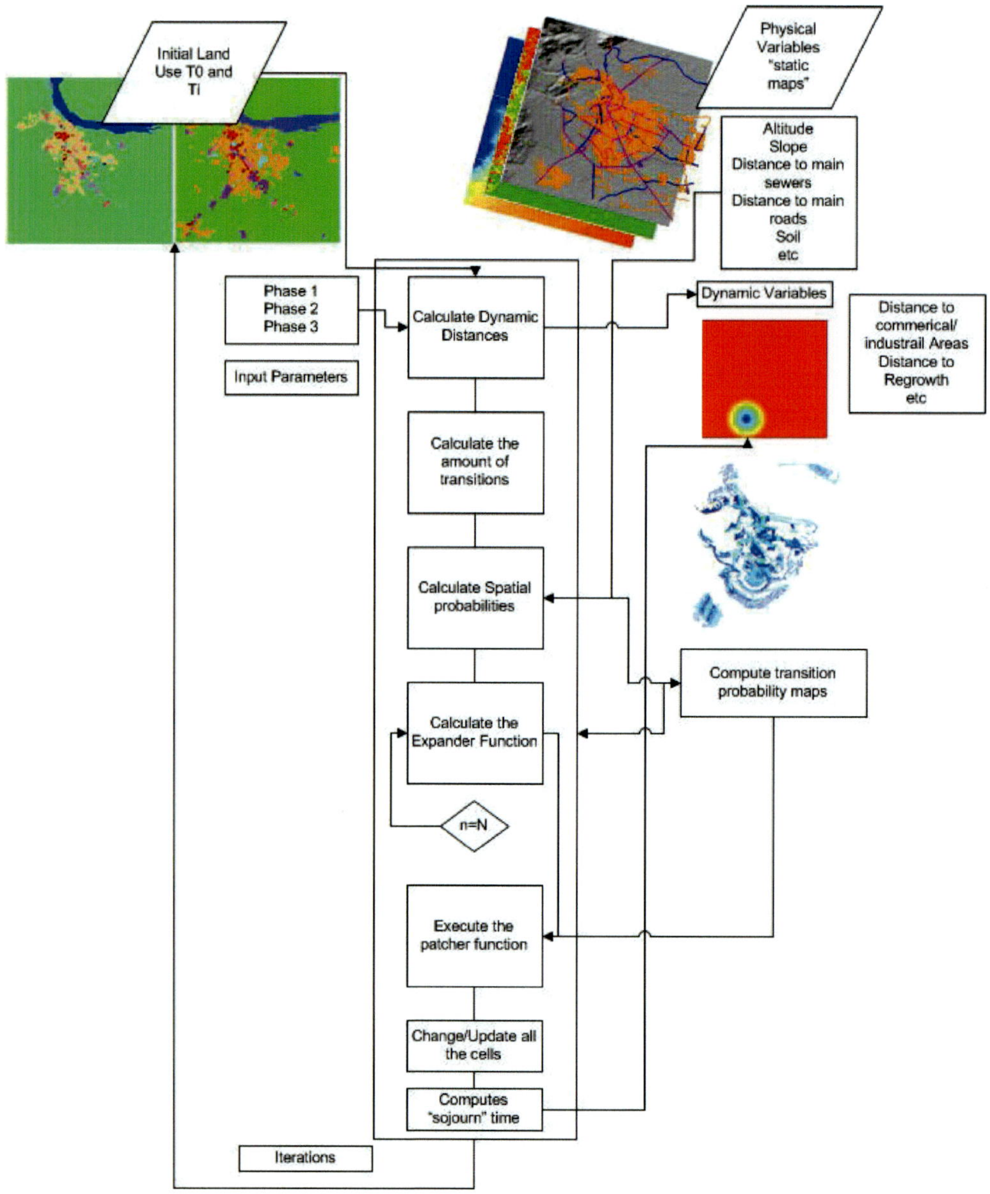

Figure 3.2. Steps to build the land use change model in dinamica

3.2.4 Validation

Because the quantity of changes is passed as a fixed parameter to the model, its validation considers only the spatial locations of the changes. This is the last procedure before the model can be used for prognosis. It consists of a comparison between the model results and a reference map, in this case, the land use map at the end of the simulation. To date, there are several map comparison techniques that have gained prominence in that they apply multiple resolution windows to assess the spatial match between two maps, e.g. Costanza (1989), Pontius (2002), Power et al. (2001) and Hagen (2003). Nevertheless, there is neither consensus about which technique yields the most appropriate validation, nor what fitness value should be taken as a threshold to accept or reject the model. Of these techniques, the fuzzy comparison method by Hagen (2003) was adapted to be used in Dinamica, named the "Reciprocal Similarity". This method employs a decay exponential function with the distance to weight the cell state distribution around a central cell. Generally, one can say that a simulated map presents good result when it has a fitness value higher than the one obtained through a direct comparison of the final and initial maps. (Hagen 2003).

Dinamica EGO contains a model to calculate the similarity maps. This model receives as inputs, the initial, the final and the simulated land use maps, and then it calculates the differences between the simulated map and the final map, based on the initial map. The comparison of similarities between the two maps (i.e. simulated and final) was carried out based on a map of differences due to the fact that simulated maps always inherit spatial patterns of the initial landscape map. Applying this technique it is possible to minimize this inheritance

This method of comparing maps always results in two values of similarity. It is advisable to choose the smaller similarity value (Soares-Filho et al. 2009). This model does not compare the changes in a cell by cell approach, but it does the comparison within a neighbourhood that is defined as a windows size. The size of the window needs to be an odd number. For the purpose of this research a window size of 11x11 was selected.

This method can be used in two ways, one that uses an exponential decay function. If the decay function is activated, it means that the importance (relative weight) of cells around the central cell decreased with the distance. If this parameter is not used, then all cells in the window size will have the same importance to calculate the similarity. Figure 3.3 shows the schematization of the fitness calculation.

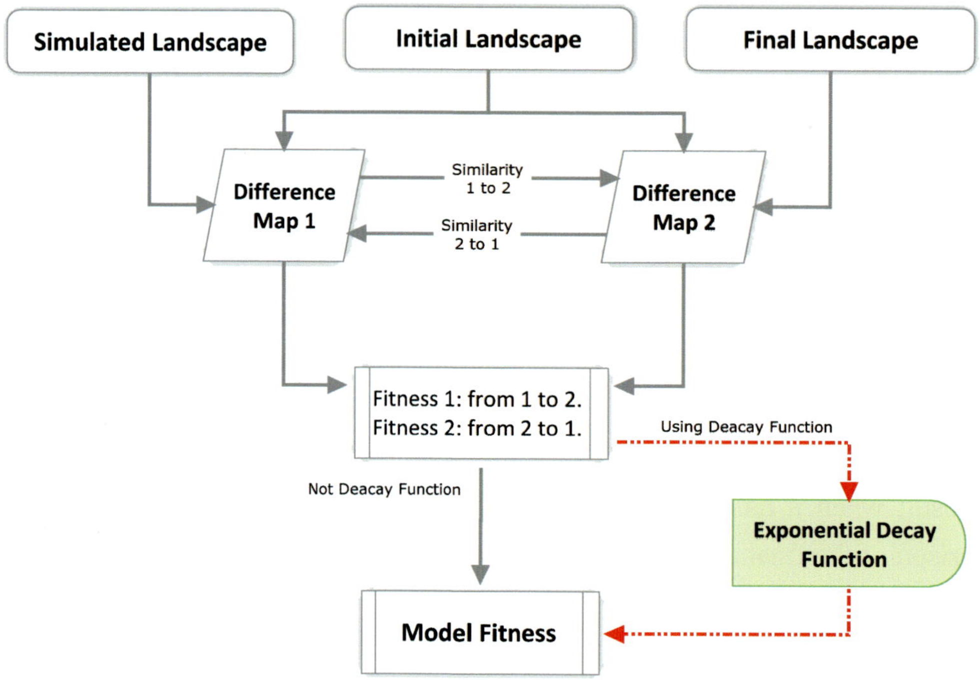

Figure 3.3 Model Fitness Conceptualization

3.2.5 Model Configuration

Figure 3.4 illustrates the land use change model configuration in Dinamica EGO. The description of the model inputs and procedures is also presented.

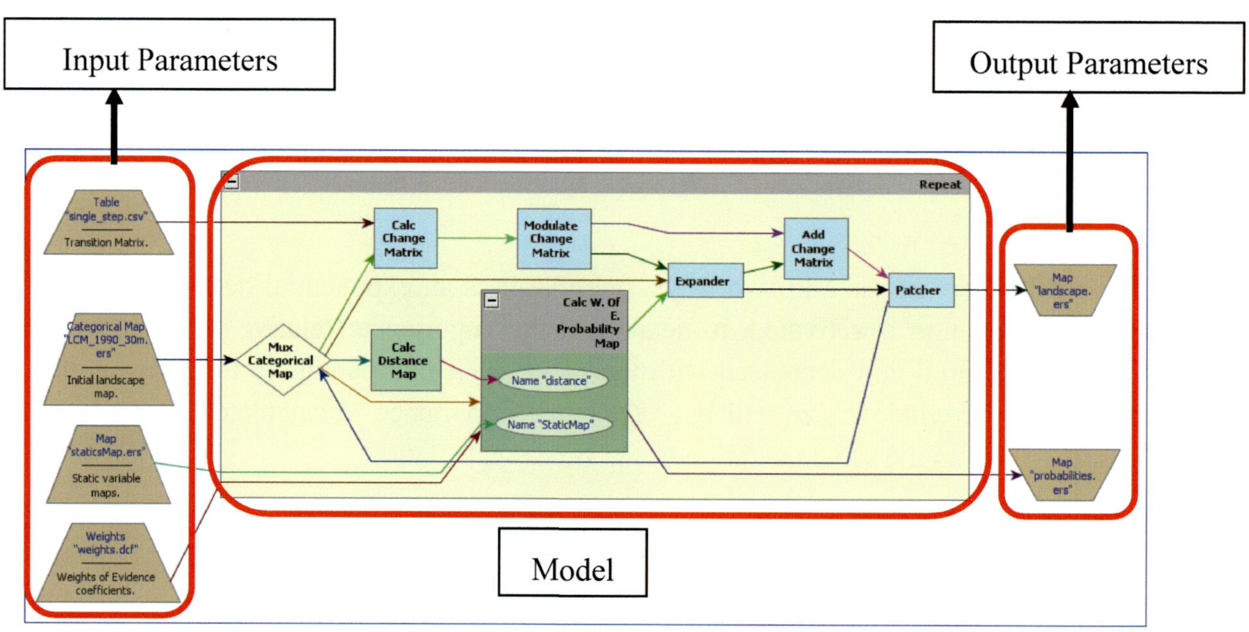

Figure 3.4 Dinamica Model for Land Use change Simulation.

✓ Input Parameters:
- Transition Matrix.
- Weights of Evidence.
- Static Variable Maps.
- Initial Land Landscape Map.

Model: Consist of a set of functions called "functors" that perform the simulation of land use change. Those functors are briefly describe here below:

- Mux Categorical Map: This functor enables dynamic update of the input landscape map.
- Calc Distance Map: Calculates a map representing the frontage distance (in meters) between a cell and the closest cell of a certain category specified in the "CategoricalMap".
- Calc Change Matrix: Receives the transition matrix, composed of net rates, and uses it to calculate crude rates in terms of quantity of cells to be changed by multiplying the transition rates by the number of cells available for a specific change.
- Modulate Change Matrix: This functor is used to split the quantity of changes between different types of transition functions, in this model between the Expander versus Patcher.
- Calc W. of E. Probability Map: calculates a transition probability map for each specified transition by summing the Weights of Evidence.
- Expander: Dedicated only to the expansion or contraction of previous patches of a certain class
- Patcher: Designed to generate or form new patches through a seeding mechanism.

Output Parameters:
- Landscape: Correspond to the simulated land use Map
- Probabilities: Output raster map that represents the probability of change.

The Expander and Patcher functors required the definition of three parameters, Mean patch Size, Variance Patch Size and Isometry. By manipulating these input parameters, the model can perform the formation of a variety of sizes and shapes of patches of change. The first two parameters are related to the size of the patches, and are set according to a lognormal probability distribution. Therefore, it is necessary to specify the parameters of this distribution represented by the mean and variance of the patch sizes to be formed, while the Isometry is related with the form of the patches to be formed.

There are not specific values that can be assigned to these set of parameters, they change from case to case and are refer as calibration parameters. The values offered by

Dinamica EGO as default were used to start any simulation. (S. Soares-Filho et al. 2009).

- ✓ Modulate Change Matrix: 0.2
- ✓ Mean patch Size: The value in hectares of 4 cells.
- ✓ Variance Patch Size: The value in hectares of 8 cells.
- ✓ Isometry: varies from 0 to 2. The patches assume a more isometric form as this number increases, a value of 1 is assumed as default.

Once all inputs are ready and the default values were assigned, the model is ready to be run for a test simulation to verify that there are not warning messages or errors.

3.2.6 Data Requirements

The application of the developed techniques requires a dataset that involves collection of data for land use/cover from at least two different years in order to determine the changes in different land use classes. Other datasets required include ancillary maps that provide data to explain the process of urbanization or land use change, such as elevation, slope, boundaries, road network, or other services including gas, electricity and cable television, rivers, water bodies, etc. Depending on the availability of maps showing plans for future housing developments, business and industrial investments, and road expansion, these can be used to assess future scenarios.

Some of the constraints involved in using this modelling paradigm come from the limit on the amount of data required and the type of data. The implementation of satellite and remote sensing techniques has enhanced access to spatial information globally in recent years. Some issues still remain, for example the type of sensors used and the objective of the project are dependent on the time the information was capture and therefore influence the quality of the dataset. Also the classification of land cover/use for agriculture purposes differs from that used for urbanization. In both cases the built-up area can be identified, but the land use within the built-up area requires ground-truth verification points for instance commercial and industrial areas.

Other issues are related to the fact that by using remote sensors the available datasets refer entirely to the physical world. In other words, is possible to have information about the environment, boundaries and some infrastructure but it is not possible to map the decisions taken by political actors, decision makers, institutions, the economists, the business market, etc. All of which can be important in setting the rules that drive urbanization. The work presented here makes use of information and datasets that are currently freely available on the Internet together with other datasets that are acquired specifically for the case studies.

For example, in the case of urban areas in Europe land use maps can be obtained from the European Environmental Agency corresponding to the Corine Dataset. The dataset exists in different resolution for the years 1990, 2000 and 2006. Since the focus of the study is on urban areas a higher level of detail is considered necessary for artificial areas and therefore the Corine land use classes were regrouped into 6 classes mainly bringing together the agriculture areas, forests and pastures into a class referred to as 'vacant land'.

The other classes inside the urban agglomeration are: the continuous urban fabric which was renamed 'residential 1', the discontinuous urban fabric was renamed as 'residential 2'; and the industrial/commercial fabric was kept the same together with recreational land use. Airports, water bodies and construction sites were separate Corine land use classes and were regrouped into a single class called 'not modelled'. They were considered static classes in this study, that is, they did not change during the analysis.

Two different sources were used for the terrain data: Shuttle Radar Topography Mission (SRTM) data with 100 meter resolution and ASTER Global Digital Elevation Model (ASTER GDEM) data with 30 meter resolution. Both sources were used to match the resolution of the land uses acquired with the Corine Data. Both DTMs were projected and used to generate the slope raster maps for 100 and 30 meter resolution.

Other datasets representing the motorways, trunks, primary and secondary roads, rails, rivers and canals were downloaded from the open street map project (http://www.openstreetmap.org/). Table 3.1 presents the datasets and the data sources used in this study.

Table 3.1 Datasets and Data sources

Group	Name / Type	Format	Data Source
Land Use Change	Roads (Primary and secondary, Motorways and Trunks)	Shape	Open Street Maps project: http://www.openstreetmap.org/ or http://download.geofabrik.de
	Rail Network		
	Rivers and Canals		
	Digital Terrain model (DTM)	Raster	SRTM (90 m Resolution): http://srtm.csi.cgiar.org Aster (30 m Resolution): http://gdem.ersdac.jspacesystems.or.jp
	Slope	Raster	Derived from DTM.
	Satellite Images	Raster	Landsat Imagery: http://glovis.usgs.gov/
	Drainage Network	Shape	Water utility company, local government
	Water Distribution Network	Shape	Water utility company, local government
	Land Cover / Land Use	Raster	Corine Land Cover (100 m Resolution): http://www.eea.europa.eu/data-and-maps Land Cover Maps - Centre for Ecology and Hydrology - CEH (25 m Resolution), Paid Information.[1] http://www.ceh.ac.uk
	High Resolution Imagery	Raster	Google Maps https://maps.google.com/

[1] License: CEH Wallingford Spatial Data License Reference number: LCM 2011-307.

3.2.7 Model Calibration

The calibration of the land use model was conducted in two phases as suggested by Almeida 2003 and Soares et al 2002. The first step consisted in the adjustment of the Weight of evidence tables that are generated by the land use change model. Those are the main drivers for the dynamics of the process. The second step consisted in the adjustment of the parameters for the mechanism of allocating the land use transitions, those correspond to the parameters of the function Patcher and Expander.

The calibration of the model is formulated as an optimization problem. The objective function used for this purpose is the fitness value computed with the reciprocal similarity method described in the section 1.2.4. The optimization is done by coupling DINAMICA EGO with the NSGA II and NSGA XP algorithm.

The coupling between Dinamica EGO and the NSGA II algorithm involved the creation of intermediary routines to handle the automatic update of the model with new parameters and to pass the objective function back to the optimizer. For the first step of the calibration, the table with the weights of evidence is manipulated. This is done following the description of the heuristic calibration of models given in the Dinamica EGO guidebook (Soares et al, 2009). According with previous results obtained by the developers of the software using different GA's algorithms and parameters, the best results are achieved when a band or interval of ± 1.2 around the initially computed values for each spatial relationship and variable is used. Figure 3.5 shows one example of a weight of evidence graph and the bounds.

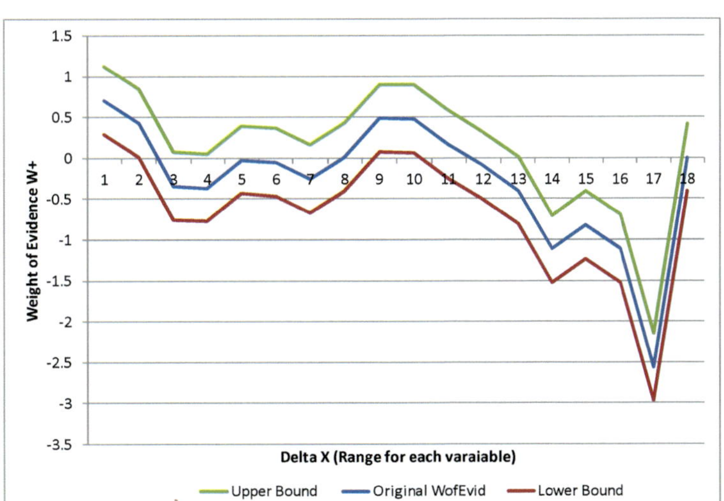

Figure 3.5 Example of Weight of Evidence graph with bounds

The optimization of the weights of evidence is performed for fixed number of iterations, depending on each model and the best set obtained in the process is used and fixed for the second step of the calibration process.

Due to the uncertainty in the values required for the transition parameters in the Expander and Patcher functors (Mean patch Size, Variance Patch Size and Isometry), and the values to be used in the Modulate Change Matrix, Estimation of such values was determined through the optimization process. According with Soares et al 2002, the calibration of these set of parameters should be done one at the time. In other words, take a particular parameter, leave the others in a predefined value, and then change this parameter until a satisfactory fitness is achieved. Then fix this one and change other, and so on until all the parameters are calibrated. To undertake this part of the calibration four intermediary routines were prepared to couple Dinamica EGO with the NSGA II and are described as follow:

1- The modulate change matrix: All the values of the expander and patcher are kept constant. The NSGA II evaluates randomly values for this matrix within the range 0 to 1. This is run for a fixed number of iterations and the best set obtained is selected for this function. These parameters are then fixed and the calibration goes on to the next function.

2- The Expander: The values for the modulate change matrix were already optimized and are fixed. The parameters for the patcher are kept constant. All the parameters for the expander (mean/variance/isometry) are changing within the optimization loop to select the best set that maximizes the objective function.

3- The Patcher: The values for the modulate change matrix and the expander were already optimized and are fixed. All the parameters for the patcher (mean/variance/isometry) are changing within the optimization loop to select the best set that maximizes the objective function.

4- All parameters together: Since the modulate change matrix provides a feedback mechanism that connects the expander and the patcher functions in Dinamica EGO, it was decided to test the calibration process in a loop where all the parameters involved in the allocation of the transitions were adjusted simultaneously. In this simulation the combination of parameters for the modulate change matrix, the expander and the patcher are changing for a fixed number of iterations. The best set that maximizes the objective function is selected.

The optimization problem was undertaken through parallel computing, with a network consisting of 5 laptops (Intel core i5, 2.4 GHz, 4 GB RAM in a 32 bit operating system). The parallel scheme was implemented in order to reduce the computational time, since it was unknown how long it will take to obtain a good set. Figure 3.6 shows the overall optimization loop used for calibration.

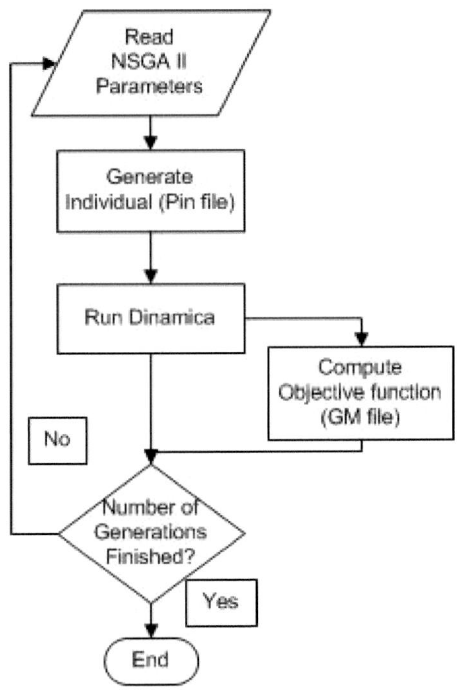

Figure 3.6 Optimization loop used for calibration

The NSGA-II algorithm was set up with the following values:

- ✓ Population size: Variable according with each model.
- ✓ Number of Generation: Variable according with the model.
- ✓ Number of objective functions: Two objectives; even though we only have one objective (Fitness or Similarity), because the nature of the optimizer chosen (Multi-objective function), it was necessary to establish as main objective the value of the fitness and as a second objective 1 - Fitness.
- ✓ Number of constrains: Zero.
- ✓ Number of Real variables: Variable according with the model .
- ✓ Lower and upper limits for variables: Variable according with the model.
- ✓ Probability of crossover: 0.9
- ✓ Probability of mutation: (1/number of variables)

- ✓ Value of distribution index for crossover: 15
- ✓ Value of distribution index for mutation: 20
- ✓ Number of binary variables: Zero.
- ✓ Plot option (0/1): 1 (Allow Plot)
- ✓ Objective function to be plotted in X axis: Objective 1 (Fitness)
- ✓ Objective function to be plotted in Y axis: Objective 2 (1-Fitness)

3.3 Case Study 1 Villavicencio, Colombia

The city of Villavicencio is a medium size municipality located in the south-east part of Colombia with an approximate population of 400.000 inhabitants in 2005, according with the data reported by the national statistics center, DANE, in 2005. Figure 3.7 shows the location of Villavicencio in Colombia. The city is located 90 Kilometres Southeast of the Bogota, the capital of the country. The average travel time is about 2 hours by car. Villavicencio is the capital city of the Meta department and is the biggest city in the east lowlands area of Colombia. Due to its geographic position it is considered as the entrance door to Bogota.

Figure 3.7 Location of the case study Villavicencio

Villavicencio has become an important commercial center, since the main roads and highways to Bogota pass through the city. Therefore, it has an important role for the economy of the whole region, serving as a transit point to Bogota and important exchange flow of goods, mainly from the agro-industry and cattle. Villavicencio and the

Meta Department have an important role in the production of natural gas and oil for the country.

The water services have a good coverage in the urban area, for both water distribution and urban drainage the level of coverage is greater than 90% DANE, 2005. The water distribution network works by gravity and supply treated water to the population. The drainage network is a combined sewerage system that discharge combine sewerage overflows to the rivers located nearby without treatment, affecting the environment and the water quality of the rivers.

Currently, there not clear figures about the economic growth of the city. Due to its geographic position and as a consequence of the internal conflict that has affected Colombia for the last 50 years, Villavicencio has experienced a high growth of population. The population growth has been mainly from people displaced from rural areas and other small municipalities of the region. This contributed to an urban development that is disperse and chaotic, completely unplanned and without controls from the local authorities. The tendency of the new developments is to be the type of slums, located along the rivers and within areas of high risk for disasters (flooding, landslides) Sanabria, 2008.

3.3.1 Data Collection

For this case a set of land use maps for the urban area were found available on the Internet as part of the tutorial exercises of the ILWIS Platform. The land use maps were produced in a framework of cooperation that existed at that time with the Colombia institute for Geographic studies IGAC. The maps correspond to the land use maps classification for the years 1960, 1978 and 1991.

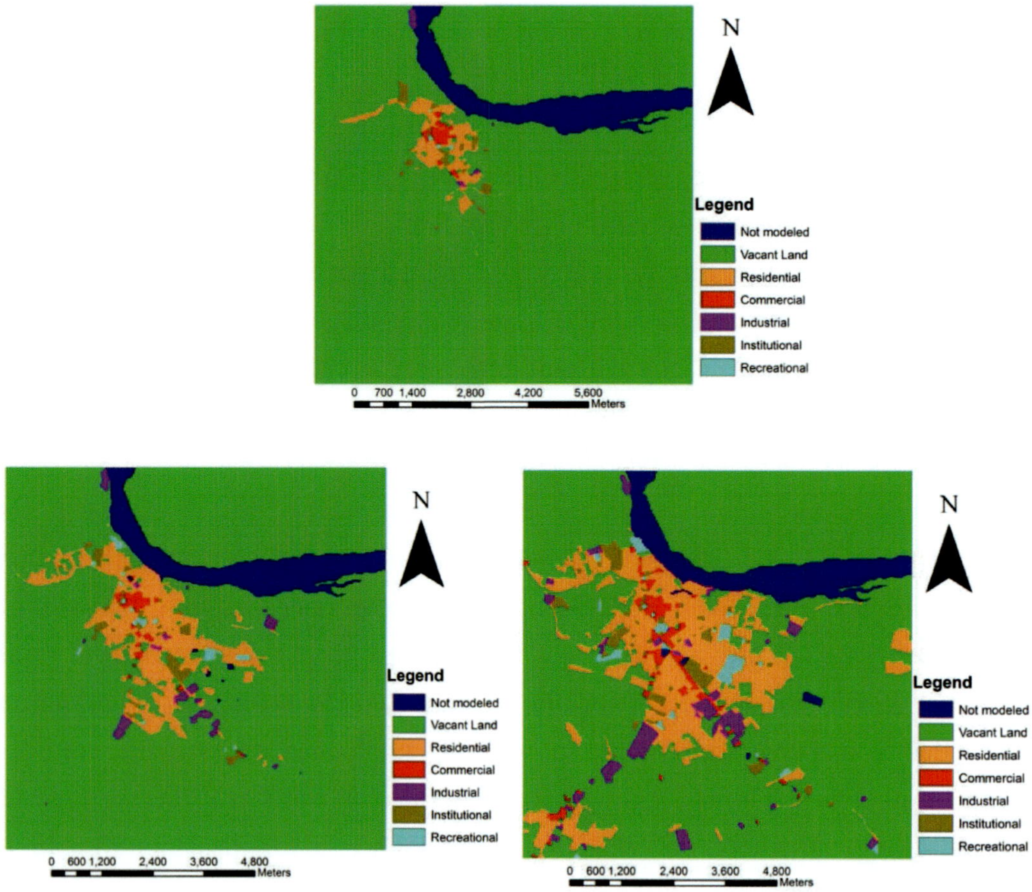

Figure 3.8 Land use maps for the years 1960 (above), 1978(left) and 1991 (right)

According to Fierro, 2007 and the land use map for the year 1960 the development of the city was along the main roads, with a clear a linear form. In 1978, it is clear that the city started to be more compact or dense around the city center and the main commercial zone. This trend continue until 1991 and probably until the present situation, with some new clusters of residential areas formed at the periphery of the main urban core, but still a long the main roads.

As it was discussed by Sanabria, 2008 at that time Villavicencio did not have a good geo-referenced cartographic database with basic information about the municipality, which is considered essential to understand the urbanization changes. Nevertheless the available information was processed to build a land use change model of the city.

A spatial dataset was collected for the study area. The dataset consisted of following maps: elevation (DTM), derived slope map, main roads in 1991 and the shape file with the lines of the main water distribution network. For the Digital Terrain Model (DTM) the Shuttle Radar Topography Mission (SRTM) data with 100 meters resolution was used. The DTM was re-projected and clipped into the proper extension of the study area. The DTM was used to generate the slope raster map. The pre-processing of the data set to build the different required maps was done with ArcGis 9.3. The water distribution model was built in Epanet 2.0 and consisted of 4100 pipes and 2800 nodes.

The cellular automata model uses this set of maps to calculate the rules of attraction or repulsion by using the Weights of Evidence method. Figure 3.9 presents the spatial dataset used to develop the land use model.

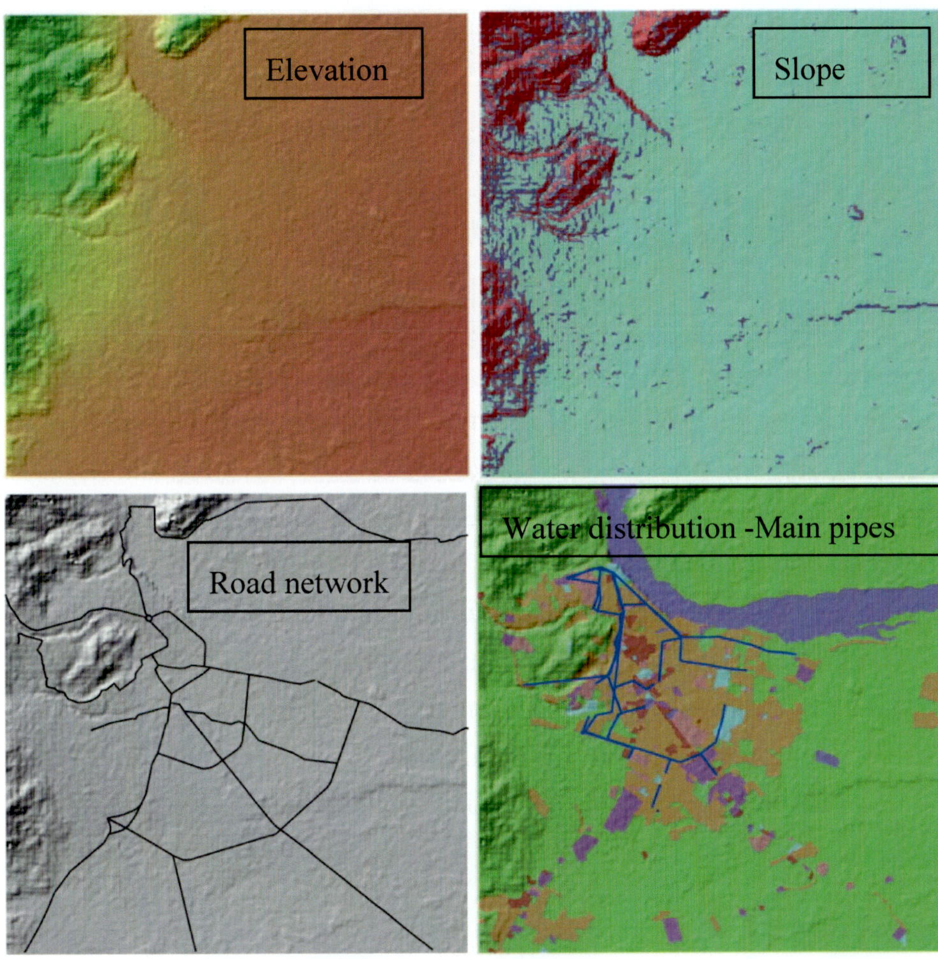

Figure 3.9 Spatial dataset used for Villavicencio

3.3.2 Initial Run

The cellular automata model was built using the land use map of 1978 and 1991 covering exactly the same area. The maps were classified into 6 classes following the results found with the corridor analysis (vacant land, residential, commercial, industrial, institutional and recreational).

The cellular automata model uses this set of maps to calculate the rules of attraction or repulsion by using the Weights of Evidence method. This results in probability a map that identifies the areas with the higher potential to be urbanized. To collect the required dataset to build this type of model is time consuming, moreover in many cities the required information does not exist or has not been collected. Figure 3.10 shows the result of the cellular automata model with the calibration of the weights of evidence matrix and the default parameters for the expander and the patcher.

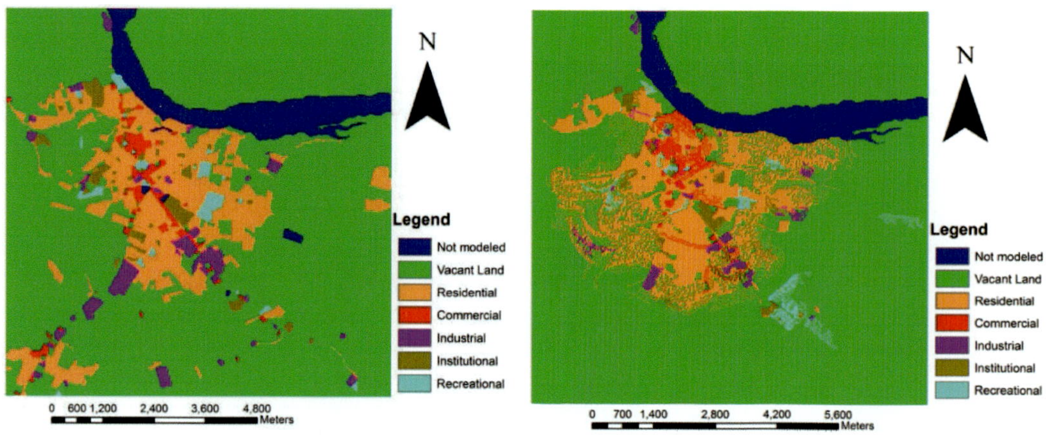

Figure 3.10 Land use map for year 1991 (left) and initial simulation for 1991 (right)

The fitness indicator for this simulation is 0.36 to further improve the model performance the calibration of the parameters for the functions expander and patcher were optimized. The best Weight of evidence matrix was then fixed to perform the following experiment:1.) The modulate change matrix (20 Variables) 2.) The Expander (60 Variables), 3.) The Patcher (60 Variables). The total number of variables for this experiment was 140 and the model was run for 8000 combinations.

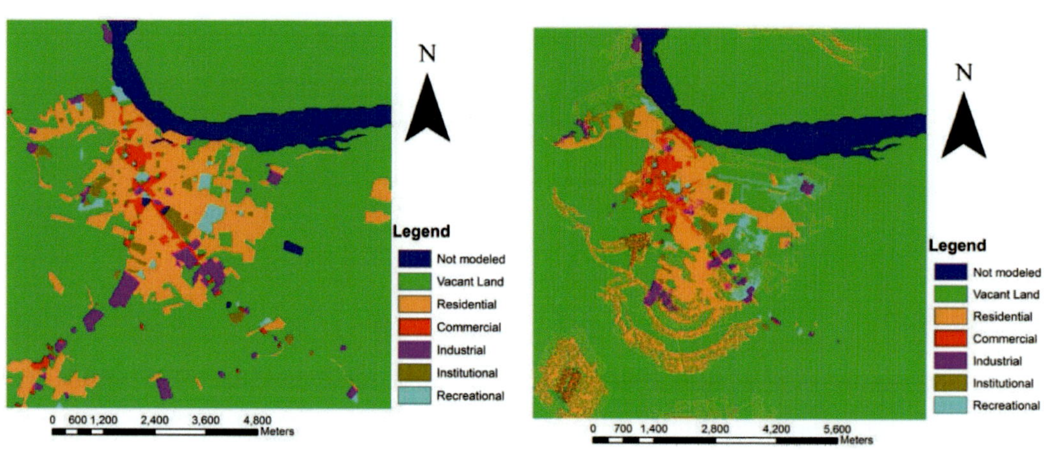

Figure 3.11 Land use map for year 1991 (left) and calibrated simulation for 1991 (right)

The calibration of the parameters expander and patcher after 8000 simulations a fitness indicator of 0.44 was achieved. As can be observed from the simulated land use map for the year 1991 presented in figure 3.11, the model is predicting a new cluster of residential area in the south west area of the city, which happen in reality but the cluster is not well structured. Moreover, big differences in the spatial allocation of all the classes can be observed despite the better fitness indicator. This shows that even tough adjusting the parameters of the model by using the calibration process it is important to have the related maps or variables that have a stronger influence in the dynamics of the land use changes. This is true for the case of Villavicencio which responds more to an unplanned urban development strategy as well as for the case of Birmingham in the UK, where there are more control from the local authorities.

3.4 Case Study 2 Birmingham, UK

The city of Birmingham is located in the West Midlands of England (see Figure 3.12). The city has an area of 267.77 km^2. It has a population of 1,028,700 inhabitants according to the City Council estimate made in 2008 (Birmingham City Council, 2012) and it forms a part of the larger West Midlands conurbation that includes other neighbouring towns such as Solihull, Wolverhampton and the towns of the Black Country. The West Midlands is the United Kingdom's second most populous urban area with a population of 2,284,093 according to the census of 2001 (Birmingham City Council, 2012). Industrial activity in the city has declined over the past fifty years. The economic crises in the 1980s caused a decrease in population that lasted until the year 2000. Since 2001 the city has experienced a rise in population as the rate of increase in population has been 4.2% between 2000 and 2010 (Birmingham City Council, 2012).

Figure 3.12. Location of the case study area

Birmingham relies on a centrally managed water supply and wastewater/sewage collection service. It has privately operated water supply and drainage/sewerage networks both managed by Severn Trent Water (www.stwater.co.uk). Much of this infrastructure dates back to Birmingham's industrial development during the 19th century, yet remains largely operational today (Darthe et al., 2008).

Population growth is leading to an increased demand for potable water and there is a corresponding increase in the flows and loads in the sewerage network and wastewater treatment plants. It is anticipated that the West Midlands population will increase by 6.6% between 2003 and 2023. This potentially poses a major problem for Birmingham because much of the area is drained by combined (stormwater and wastewater) collector systems and as such there will be an increased need to control runoff from rainfall events and attenuate localised flood risk (Last et al., 2011).

3.4.1 Data collection

For this case the land use maps corresponding to the Corine Dataset from the European Spatial Agency were used. The maps correspond to the land use classification for the years 1990, 2000 and 2006. Because the focus of the study is the urban areas a higher level of detail is considered for artificial areas, therefore, the Corine land use classes were regrouped into 6 classes mainly grouping the agriculture areas, forest, pastures etc, into a class call vacant land. The other classes inside the urban agglomeration were kept equal. Figure 3.13 shows the different land use maps.

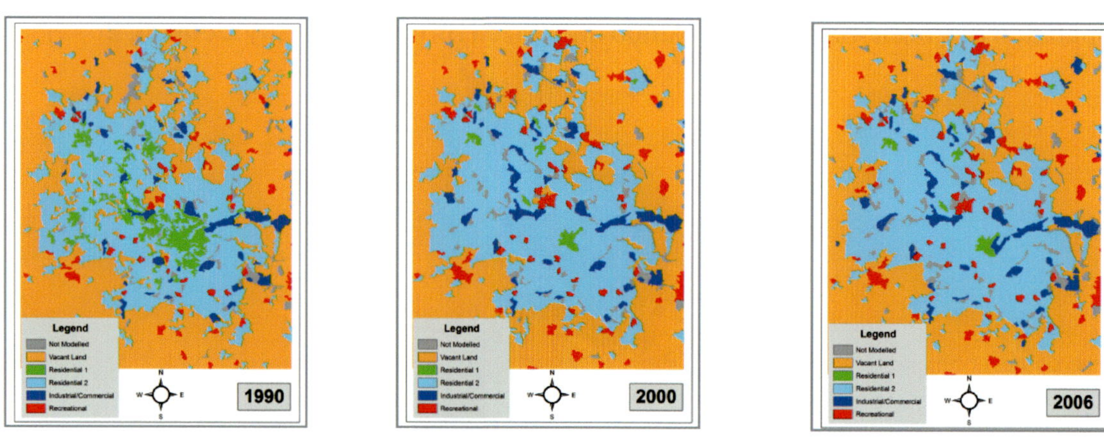

Figure 3.13 Corine Land cover Map, for the years 1990, 2000 and 2006

A spatial dataset was collected for the study area. The dataset consisted of following maps: elevation (DTM), derived slope map, main trunks and motorways, primary and secondary roads, rail network, main rivers and canals, and the shape file with the lines of the drainage network. For the Digital Terrain Model (DTM) two different sources were used: Shuttle Radar Topography Mission (SRTM), data with 100 meters resolution and ASTER Global Digital Elevation Model (ASTER GDEM), data with 30 meters resolution, both DTM's were re-projected and clipped into the proper extension of the study area. The DTM was used to generate the slope raster map.

The shape file representing the motorways, trunks, primary and secondary roads, rails, river and canals were downloaded from open street map project. (http://www.openstreetmap.org/). A shape file with the lines of the urban drainage system for the region serve by Severn Trend was made available through the learning alliance established in Birmingham.

The urban drainage model used in this study was provided by Severn Trend and corresponded to the region of the Upper Rae Main catchment. Previous research done with this dataset includes Thuy, 2009 and Last, 2011. The maps are presented in figure 3.14.

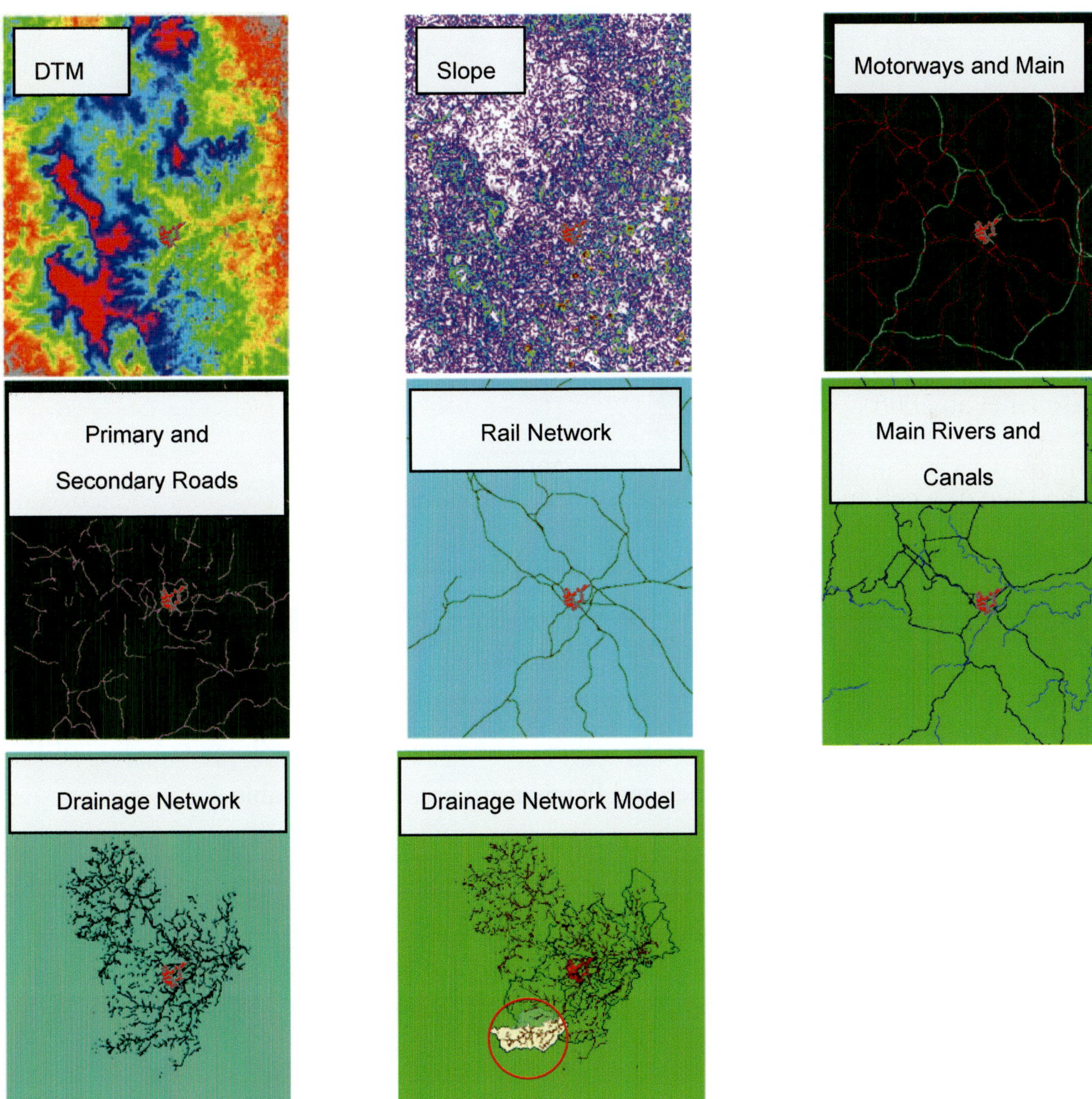

Figure 3.14. Spatial dataset used in this study.

3.4.2 Initial run

The model was run with initial land use map for the year 1990 and for time span of 10 years to simulate the land use for the year 2000.The initial simulated map is shown in figure 3.15 with comparison to the real land use map of the year 2000. To compare the output of the simulation with the real land use map Dinamica can calculate a similarity map to check the spatial correlation between the two, this initial correlation was 0.18. relatively low as can be observed from figure 3.15.

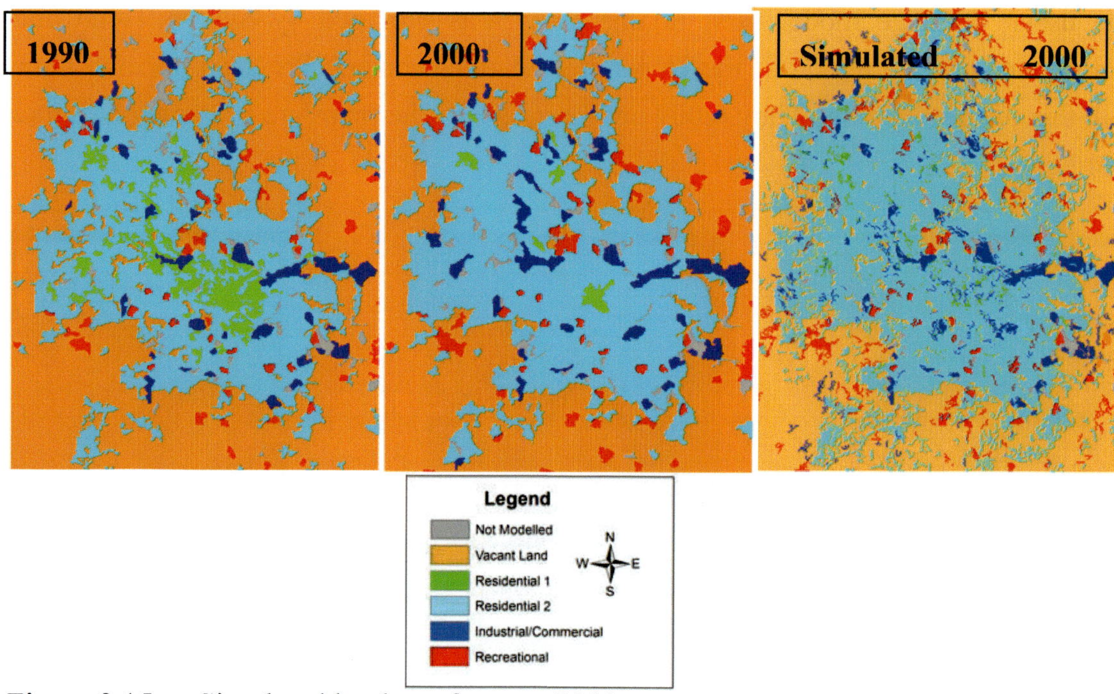

Figure 3.15. Simulated land use for year 2000.

To improve the model performance the calibration procedure described in the previous sections was followed. The first step was to adjust the Weights of Evidence matrix initially computed by Dinamica. This consisted of 150 functions that describe the attraction/repulsion effect of each variable per transition. The functions were adjusted within the bound of ± 1.2 of the initial value. The ranges used for the parameters of the Modulate change matrix, Expander and Patcher are presented in Table 3.2.

The first run was to adjust the Weights of evidence matrix. This run alone improved the fitness of the model from 0.18 obtained initially to 0.41.

Table 3.2 Variables and intervals used for Calibration

Transitions		Modulate Change Matrix (%)	Expander			Patcher		
From	To		Mean Patch Size (Ha)	Patch Size Variance (Ha)	Patch Isometry	Mean Patch Size (Ha)	Patch Size Variance (Ha)	Patch Isometry
1	3	0 - 1	1 - 2000	2 - 3000	0 - 2	1 - 2000	2 - 3000	0 - 2
1	4	0 - 1	1 - 2000	2 - 3000	0 - 2	1 - 2000	2 - 3000	0 - 2
2	3	0 - 1	1 - 2000	2 - 3000	0 - 2	1 - 2000	2 - 3000	0 - 2
2	4	0 - 1	1 - 2000	2 - 3000	0 - 2	1 - 2000	2 - 3000	0 - 2
3	1	0 - 1	1 - 2000	2 - 3000	0 - 2	1 - 2000	2 - 3000	0 - 2
3	2	0 - 1	1 - 2000	2 - 3000	0 - 2	1 - 2000	2 - 3000	0 - 2
3	4	0 - 1	1 - 2000	2 - 3000	0 - 2	1 - 2000	2 - 3000	0 - 2
4	3	0 - 1	1 - 2000	2 - 3000	0 - 2	1 - 2000	2 - 3000	0 - 2
Number of Variables for Calibration		8	24			24		

The best Weight of evidence matrix was then fixed to perform the following experiments:1.) The modulate change matrix (8 Variables, 1728 simulations), 2.) The Expander (24 Variables, 1728 Simulations), 3.) The Patcher (24 Variables, 1728 Simulations). The results of the best simulation are presented in table 3.3, as well as the best fitness value.

Table 3.3 Values for each variable and best fitness value

Transitions		Modulate Change Matrix (%)	Expander			Patcher		
From	To		Mean Patch Size (Ha)	Patch Size Variance (Ha)	Patch Isometry	Mean Patch Size (Ha)	Patch Size Variance (Ha)	Patch Isometry
1	3	0.86	231.78	1856.25	0.26	984.98	1146.50	1.13
1	4	0.71	2.36	1037.73	0.68	1493.70	2802.61	1.26
2	3	0.90	1918.89	2544.72	0.52	1417.12	783.16	0.85
2	4	0.10	1863.19	1370.60	1.63	1443.31	1026.31	1.66
3	1	0.56	481.85	1879.25	0.66	1787.29	1796.29	1.60
3	2	0.22	15.12	523.95	0.76	1345.92	231.77	1.56
3	4	0.75	945.30	1860.94	1.24	805.57	2499.00	1.55
4	3	0.93	566.09	1985.54	1.70	1461.52	870.11	0.89
Fitness of the best simulation		0.46	0.45			0.46		

The best values found in the calibration of each individual function (i.e Modulate change matrix, Expander and Patcher) were then used to build a single model with the best parameters. The value achieved of the fitness indicator was 0.42. This value of fitness is lower than the ones achieved by the calibration of each individual function. This can be explained because there is a direct relation between the amount of changes or cells that are allocated by the model with the Patcher and the Expander. Therefore, the combination of these parameters is important, rather than the adjustment of each individual sub-set of parameters per function. Another test was prepared for adjusting the parameters of the Modulate change matrix, the Patcher and the Expander at the same time. The results show similar values for the parameters of each function (Modulate change matrix, Expander, Patcher) and the value of the fitness was 0.45. Not big differences were achieved. Figure 3.16 shows the simulated map for year 2000 with the calibrated parameters.

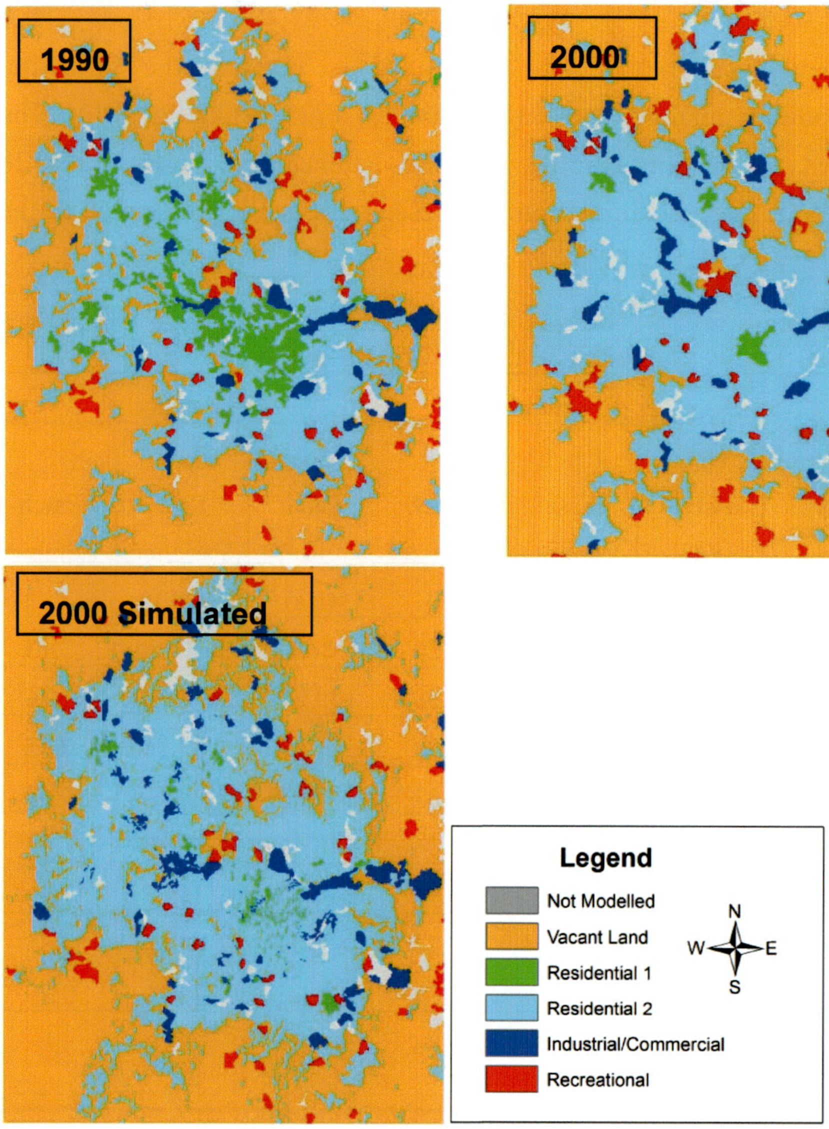

Figure 3.16. Simulated land use for year 2000 after calibration.

As presented in table 3.3, the automatic calibration process is promising because all the evaluations needed to improve the model performance are time-consuming to be done manually. Particularly, in this case the interest in developing the model is to simulate the internal changes that occur in the city. With the amount of land use classes to be modelled the case becomes more complex. The fuzzy similarity indicator to compare the simulated land use map and the real map showed a good correlation, more than 0.40. Even though, it is possible that all the variables that can explain the internal urban dynamics are not used. Moreover, it is possible that the necessary information and datasets needed to achieve higher correlation factor are probably not measured in reality. The interval of the parameters for the expander and the patcher is quite big and there are not many values reported in the literature or other studies. It is also interesting to note by observing the simulated map for year 2000 presented in figure 3.16 that the residential 1 category or high density residential is disappearing from the simulation and the model tend to allocate the cells spread in the modelling area, no big clusters are formed. This outcome is not expected in reality since this class correspond to the historical city center of Birmingham. But, observing the land use maps for the year 1990 and 2000 there is a significant decrease in the area covered by this class. Even though during the decade 1990-2000 the city of Birmingham experimented a decrease in population. The city started to gain population again around the year 2000 when the city changed from industrial city into a services oriented city. The reduction of the area for this class can also suggest that the land use maps were produced with different data sources and with different methodologies. To explore this issue more data was collected for this case, the following section describes the analysis that was performed to update the model.

3.4.3 Updating the model and second run

From the visual analysis of the reclassified maps produced by using CLC data (see figure 3.13) it was clear that the method from which this information was derived was not the same. This can be observed by analyzing the patches of land use Residential 1 (light green in the maps), where from year 1990 to year 2000 it almost disappears, and only the historical centre of Birmingham appears classified in this class for the latter. Using graphical inspection the maps from year 2000 and 2006 appear to be produced using the same approach.

For the reasons stated above, it was assessed as appropriate to compare the produced maps of land use against satellite images, in order to see whether or not the reduction in size of land use residential occurs between the two first years of comparison. The only source of satellite images with the spatial and temporal resolution extension needed to perform this analysis were Landsat images (see annex 2), this analysis was performed and presented in Medina 2012. The conclusion of this comparison was that the land use maps for the years 1990 and 2000 were produced with different methodologies, since it was not apparent that the residential 1 cluster had a significant reduction in size. With this in mind the land use map for the year 1990 was discarded and instead the model for land use changes was updated to use as initial landscape the year 2000 and the final landscape the year 2006, the resolution of the dataset is the same and corresponds to 100 meters. Another conclusion of this analysis was the need to find another source of information that includes in more detail the residential urban part of Birmingham, UK. Due to the problems founded with the 100 meter resolution land use data from EAA (Corine Land Cover), an additional model with information from the centre of Ecology and Hydrology of UK, CEH was built (Morton et al. 2011). The information of the CEH was used as primary source of information to derive the Corine dataset for the UK and it has a higher resolution (30 meters). Maps of land use, from the years 1990, 2000 and 2007 were available, but again all of them were produced using different methodologies and different purposes as it was founded after the reviewing of the final quality report of each data[2] set.

The land use data from CEH (Morton et al. 2011) was reprojected from British National Grid into ETRS_1989_LAEA, and then a resample was applied to them in order to have the desired resolution, from 25 to 30 meters. A reclassification was also performed to preserve the same classes used for the model of 100 meter resolution.

The reclassified rasters are presented in figure 3.17, 3.18 and 3.19, for the years 1990, 2000 and 2007 respectively. Observing the results of these reclassification two major issues were identified. The first one corresponds to a significant expansion of class residential 1, from the year 1990 to the year 2000, and then from the year 2000 to the year 2007 it presents a significant reduction in size. To solve this issue the Landsat images were used again to identify if this expansion and shrinking process really happened or not. The analysis shows that the exaggerated expansion of year 2000 does not corresponds to the reality, but the information from years 1990 and 2007 were well represented when comparing between the reclassified maps and the Landsat images. Therefore, the map produced for the year 2000 was discarded from the analysis.

[2] There is a final report that accompanies each data set (1990, 2000 and 2007).

The second issue is related with the reclassification itself, given the land uses in the original dataset provided by CEH, it was not possible to differentiate the land uses Industrial/Commercial and the land use Recreational as it was done with the initial model that uses the Corine Dataset.

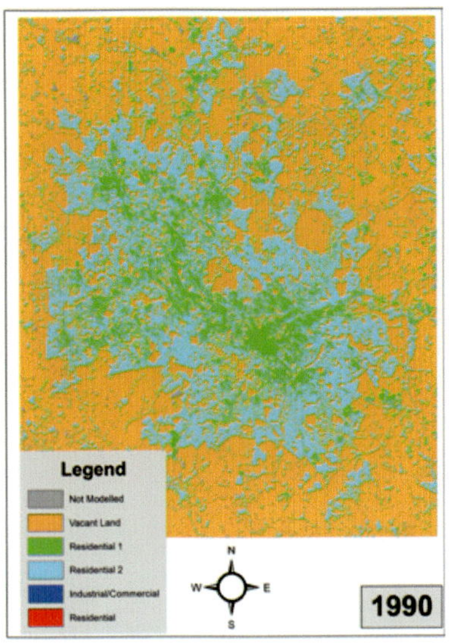

Figure 3.17 CEH Map, Year 1990 (Model M1 -30m).

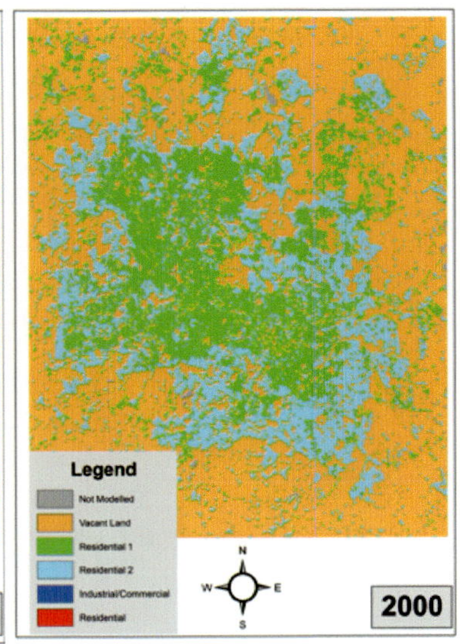

Figure 3.18 CEH Map, Year 2000 (Model M1 -30m).

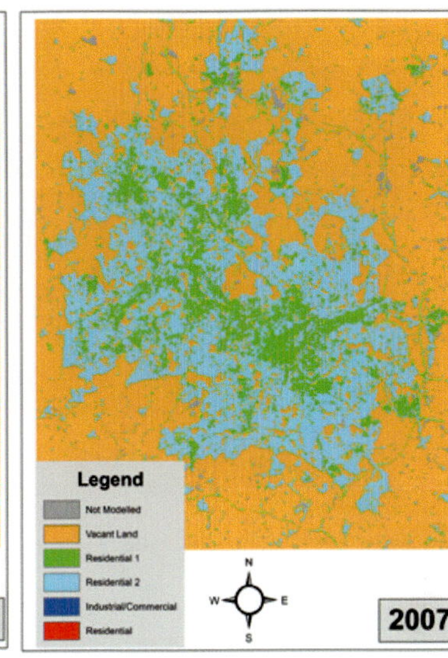

Figure 3.19 CEH Map, Year 2007 (Model M1 -30m).

To solve the second issue, it was decided to combine the information from CEH and the information from EEA. To do so, the main criterion to combine both raster was to preserve as much as possible the integrity of the CEH data (given the higher resolution of it). When combining the rasters using the Mosaic method of ArcGIS, all the information of the raster from CEH will be assigned to the output, except in those areas where the land use in the EAA raster is either Industrial/Commercial or Recreational.

The final maps of land use to be used in the model M1 are presented in figure 3.20 and figure 3.21 As initial landscape is presented the year 1990 and the final landscape correspond to the year 2007.

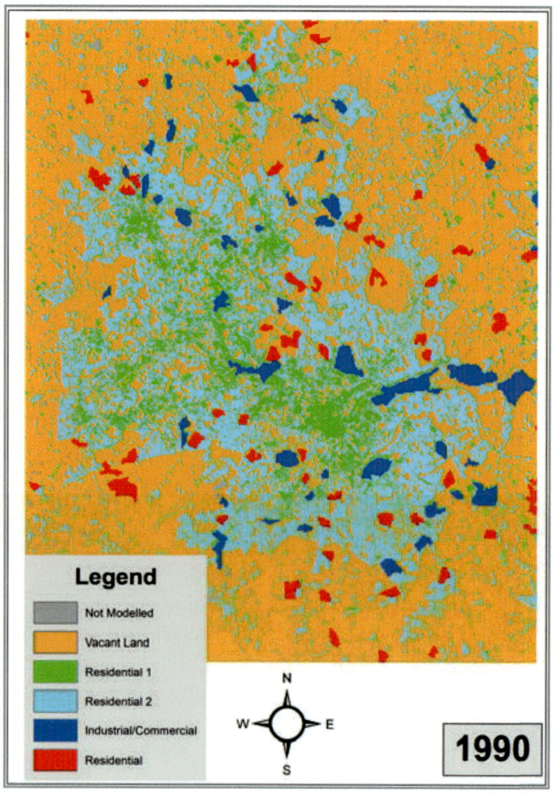

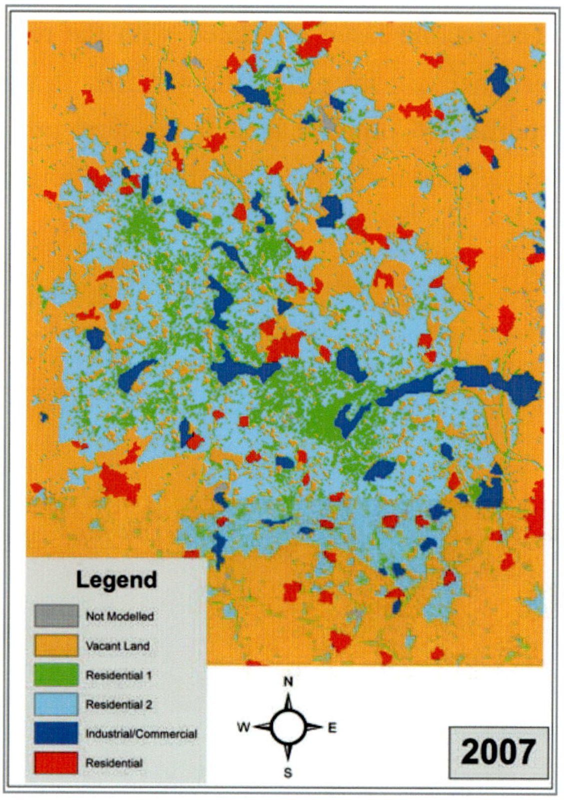

Figure 3.20. Initial Landscape model M1 (30m) Year 1990.

Figure 3.21. Final Landscape model M1 (30m) Year 2007.

3.4.4 Model Configuration

✓ **Model M1 30 meter resolution:**

Two different transition Matrices were calculated; Single step and Multiple step, the first one correspond to the total rate of changes between the initial landscape (1990) and the final landscape (2007), while the second one corresponds to the rate of change on a year basis. These matrices are presented in table 3.4and table 3.5 respectively.

The transition matrices show that there will be changes from urban soil (classes 2, 3, 4 and 5) to vacant land; but in reality this kind of transitions are unlikely. In the urbanization process, it is unlikely that urbanized soil will return to vacant land, pastures, etc unless is a park or recreational area due to the added value that was already generated. Therefore, it was decided to readjust these matrices for the next steps in the modelling. All the transition to land use 1 (Vacant Land), where neglected.

Table 3.4. Single Step Matrix Model M1

Single Step Original		TO 1	2	3	4	5
FROM	1		0.0173	0.1084	0.0107	0.0290
	2	0.1256		0.2880	0.1204	0.0055
	3	0.2168	0.0815		0.0193	0.0060
	4	0.0669	0.0313	0.0945		0.0011
	5	0.0887	0.0078	0.0716	0.0052	

Table 3.5. Multiple Step Matrix Model M1

Multiple Step Original		TO 1	2	3	4	5
FROM	1		0.0010	0.0082	0.0006	0.0020
	2	0.0072		0.0287	0.0110	0.0003
	3	0.0165	0.0083		0.0008	0.0002
	4	0.0038	0.0024	0.0067		0.0056
	5	0.0056	0.0003	0.0051	0.0003	

In the original transition matrices 20 possible transitions are represented, while in the simplified matrices only 16 changes are possible. Analyzing the transition rates obtained in the transition matrices, it can be observed a tendency of land use Residential 1 to be transformed into land uses residential 2 and into land use Industrial/Commercial. This situation is confirmed by the literature reported by SWITCH project (European Commission, 2010) showing that the city of Birmingham is changing from a industrialized city into a service oriented urban area.

✓ **Model M2 100 meter resolution:**

Two different transition matrices were generated for model M2, single step and multiple step, the first one correspond to the total rate of changes between the initial landscape (2000) and the final landscape (2006), while the second one corresponds to the rate of change in a year basis. These matrices are presented in table 3.6 and table 3.7 respectively.

Following the same arguments discussed for model M1, the transitions from urbanized soil to vacant land are neglected.

Table 3.6 Single Step Matrix Model M2

Single Step		TO				
Original		1	2	3	4	5
FROM	1		-----	0.0142	0.0037	0.0080
	2	-----		0.0164	0.0019	-----
	3	0.0256	0.0003		0.0078	0.0014
	4	0.0243	-----	0.0183		0.0002
	5	0.0350	-----	0.0171	0.0327	

Table 3.7 Multiple Step Matrix Model M2

Multiple Step		TO				
Original		1	2	3	4	5
FROM	1		-----	0.0015	0.0009	0.0015
	2	-----		0.0026	0.0004	-----
	3	0.0042	0.0001		0.0013	0.0002
	4	0.0026	-----	0.0056		0.0051
	5	0.0050	-----	0.0044	0.0054	

In the original transition matrices 16 possible transitions are represented, while in the simplified matrices only 12 changes are possible.

The transition rates obtained for both models M1 and M2 are showing the following:

- ✓ The land use Vacant land is transforming as time passes into Residential 2, this is a very typical phenomena where the surrounding areas to urban settlements start growing in the periphery of the city to allocate the incoming population (Almeida et al. 2003).
- ✓ Land use Residential 1 tends to transform its use into Industrial/Commercial which is also a common phenomenon of urban cities. The centre of the city tends to change from residential land uses into commercial settlements such as malls and commercials buildings (Almeida et al. 2003).

Correlation Analysis

Although there is no agreement on what threshold should be used to exclude a variable from the analysis, in the literature review were found some suggested values that can be used for this purpose. Bonham-Carter (1994), reports that values less than 0.5 for Cramer's Coefficient and the Joint Information Uncertainty suggest less association rather than more. Almeida et al. 2005, suggest more restrictive values, for the Cramer' s

Coefficient, the empirically established threshold value was 0.45, and for the Joint Information Uncertainty, 0.35. Using the thresholds suggested by Almeida in all comparisons made in this study, these associations are less than these threshold. As a conclusion of this analysis all the maps used in the analysis are independent among them, and can be used to construct the model of land use change as driving factors.

3.4.5 Running the model M1 and M2

To test that the models run adequately, a first run of both models M1 and M2 was executed using the default values for the parameters that controls the transitions inside Dinamica EGO. For both models the default values are presented in table 3.8.

Table 3.8. Default Values for Model M1 and M2

Parameter	Default Value
Modulate Change Matrix	0.2
Mean patch Size	0.36 ha
Variance Patch Size	0.72 ha
Isometry	1.0

Both models were initially run and the fitness function or similarity map was calculated using the two methods available: with decay function and without it. For model M1 the fitness values are 0.37 with decay function and 0.58 without it. For model M2 the fitness values are 0.19 with decay function and 0.36 without it.

3.4.6 Calibration of model M1 and M2

The initial part of the calibration corresponded to the adjustment of the weights of evidence table originally calculated by Dinamica EGO. For this purpose each model M1 and M2 was run 2240 times. This number of iterations corresponds to 224 individuals in the population and 10 generations. The result of this experiments showed that it was not possible to gain any extra percentage in the fitness functions in comparison with the ones achieved with the default values. Therefore, the original table of Weights of evidence was preserved for the other simulations.

The second part of the calibration process corresponded to the adjustment of the parameters of the expander and patcher functions that control the transition. Table 3.9 shows the total amount of variables obtained for the calibration process in both models.

Table 3.9. Initial number of variables in the calibration process

Parameter	Number of Variables	
	Model M1	**Model M2**
Expander		
Mean Patch Size	16	12
Variance Patch Size	16	12
Isometry	16	12
Patcher		
Mean Patch Size	16	12
Variance Patch Size	16	12
Isometry	16	12
Modulate Change Matrix	16	12
TOTAL	**112**	**84**

To try to simplify the optimization problem a sensitivity analysis was carried out manually. This process also helped to define the search space where the optimal values of these variables may be. The sensitivity analysis was done with a set of constant parameters while changing the values of one variable at a time (see table 3.10). The process was done running the simulation model and then reporting the Fitness of the resulting simulated map. The similarity analysis of maps or fitness was calculated using the exponential decay function and without it.

Table 3.10. Values used in the sensitivity analysis

Parameter	Constant Value	
	Model M1	**Model M2**
Mean Patch Size	0.36	4
Variance Patch Size	173	8
Isometry	1.0	1.5

The results of the sensitivity analysis for model M1 are presented in figures 3.22 to 3.27.

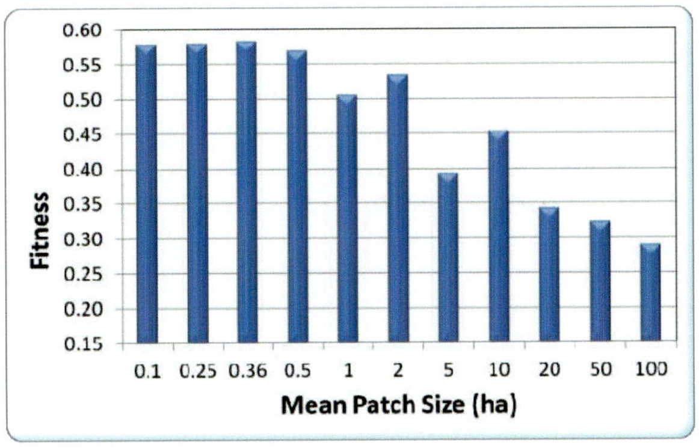

Figure 3.22 Sensitivity Analysis Mean Patch Size, Not decay Function, model M1

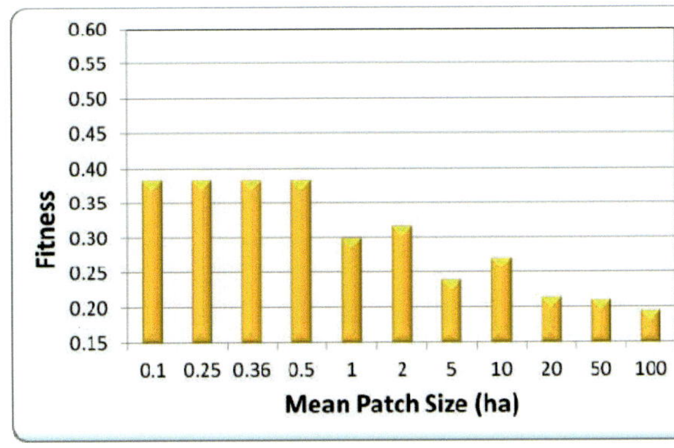

Figure 3.23 Sensitivity Analysis Mean Patch Size, With decay Function, model M1.

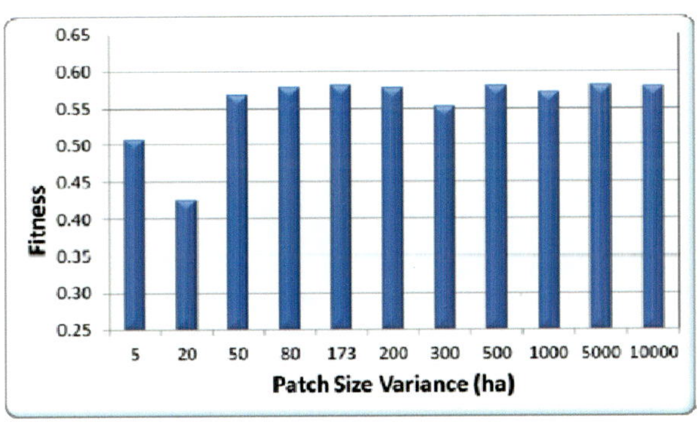

Figure 3.24 Sensitivity Analysis Variance Patch Size, Not decay Function, model M1

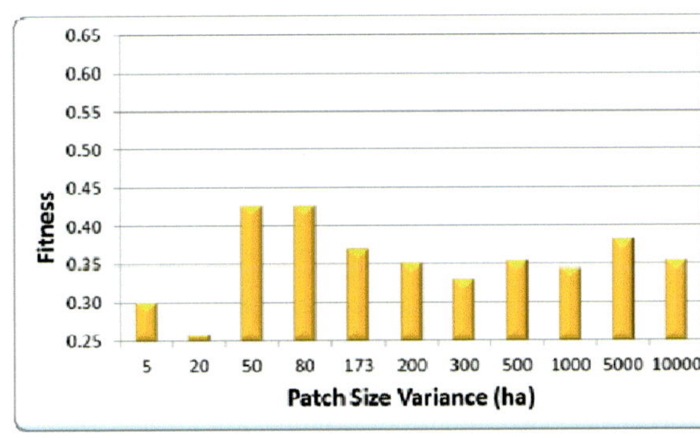

Figure 3.25 Sensitivity Analysis Variance Patch Size, With decay Function, model M1.

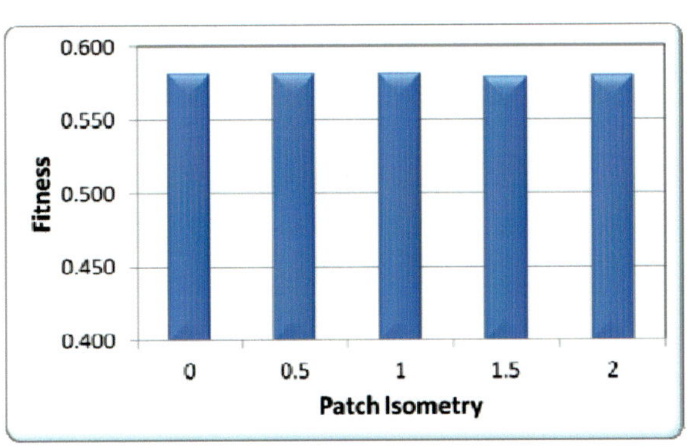

Figure 3.26 Sensitivity Analysis Isometry, Not decay Function, model M1

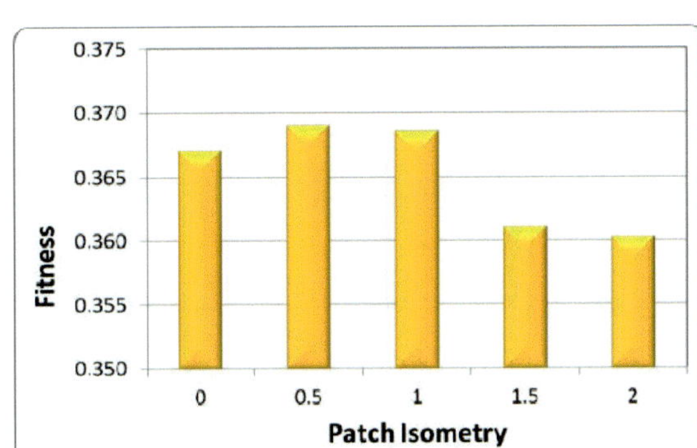

Figure 3.27 Sensitivity Analysis Isometry, With decay Function, model M1.

Similarly, the results of the sensitivity analysis for model M2 are presented in figure 3.28 to 3.33

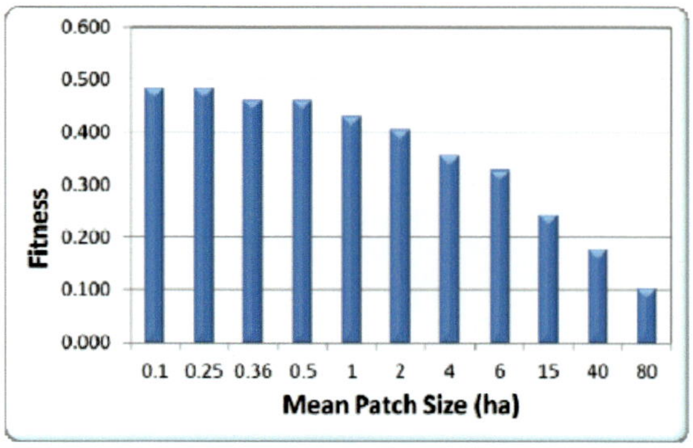

Figure 3.28 Sensitivity Analysis Mean Patch Size, Not decay Function, model M2

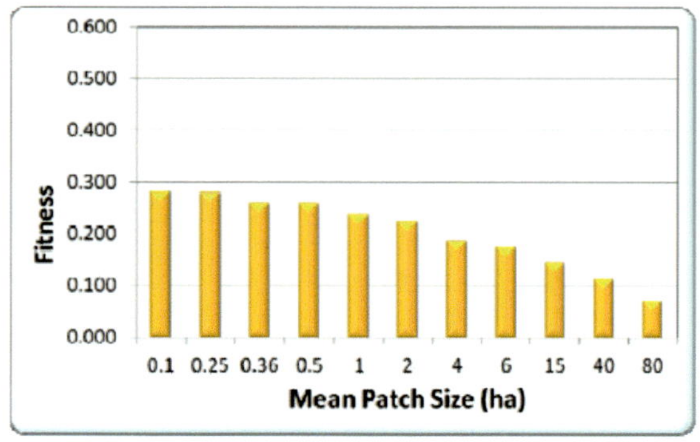

Figure 3.29 Sensitivity Analysis Mean Patc Size, With decay Function, model M2.

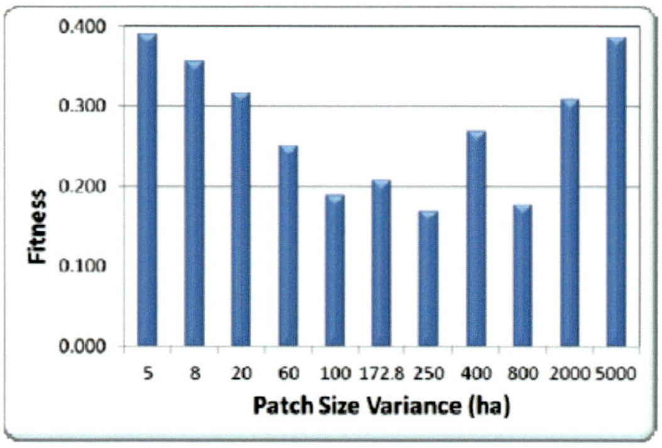

Figure 3.30 Sensitivity Analysis Variance Patch Size, Not decay Function, model M2

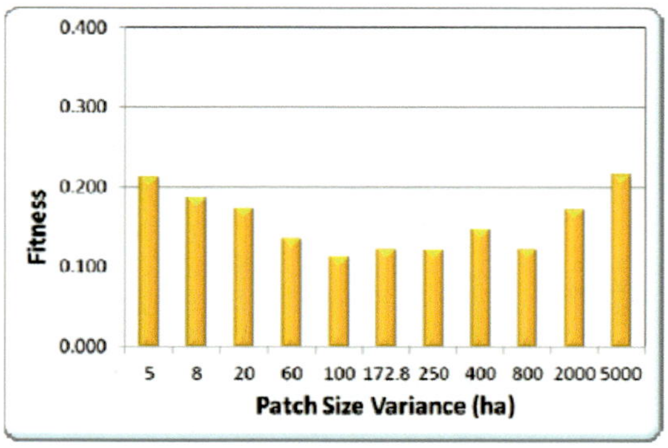

Figure 3.31 Sensitivity Analysis Variance Patch Size, With decay Function, model M2.

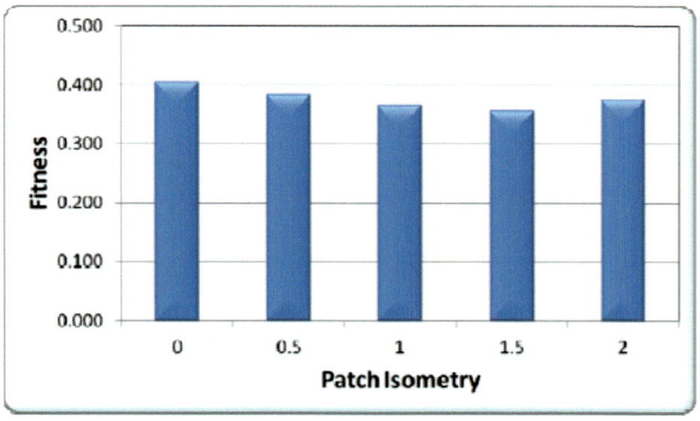

Figure 3.32 Sensitivity Analysis Isometry, Not decay Function, model M2

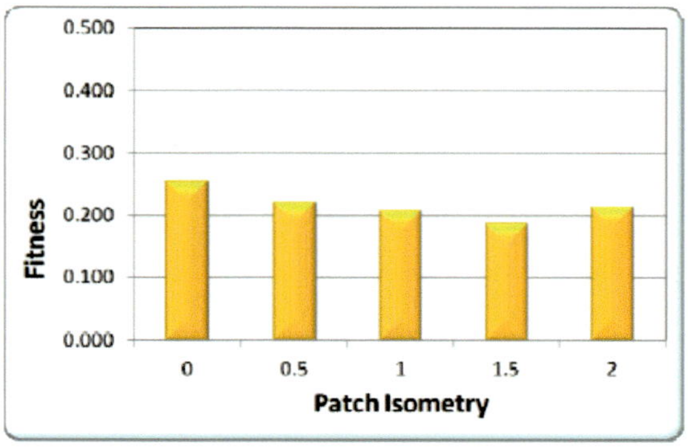

Figure 3.33 Sensitivity Analysis Isometry, With decay Function, model M2.

After this analysis a range for each of the variables was selected to be used in the calibration process, they are summarized in 3.11. Additionally it was observed that the response for parameter isometry was more or less constant for all the simulations, therefore it was fixed and excluded from the calibration process.

Table 3.11. Ranges and values of Parameters to be used in the calibration

Parameter	Range or Value	
	Model M1	**Model M2**
Expander and Patcher		
Mean Patch Size	0 - 2	0 - 2
Variance Patch Size	50 - 200	0 - 200
Isometry	1.0 (1)	1.0 (1)
Modulate Change Matrix (2)	0 - 1	0 - 1

(1) Except for all the transition to land use 1, where 0.5 was used.

(2) This corresponds to a percentage.

From these new ranges of values we selected one value for each parameter and models M1 and M2 were run again (see table 3.12).

Table 3.12. Adjusted Values of Parameters

Parameter	Adjusted Value	
	Model M1	**Model M2**
Expander and Patcher		
Mean Patch Size	0.36	1.0
Variance Patch Size	172.8	5.0
Isometry	1.0 (1)	1.0
Modulate Change Matrix	0 - 1	0 - 1

(1) Except for all the transition to land use 1, where 0.5 was used.

These set of runs were called run with adjusted values. The fitness results of the new runs were then compared with the fitness obtained using the default values presented in table 3.8. The comparison is presented in table 3.13.

Table 3.13. Model Fitness Default vs Adjusted Values

Model	Fitness			
	Model M1		**Model M2**	
	Decay Func.	**Not Decay**	**Decay Func.**	**Not Decay**
Default Values	0.3699	0.5829	0.1859	0.3560
Adjusted Values	0.3680	0.5804	0.2761	0.4728
% Improvement	-0.52	-0.44	48.54	32.82

The result presented in table 3.13 show no improvements for model M1 with the new set of parameters. While for model M2 a significant improvement is achieved.

To calibrate the remaining parameters the land use change model engine was coupled with the NSGA II. To speed up the calibration process the parallelized version of the NSGA II was used. The NSGA XP was developed by Barreto 2012. This algorithm was parallelized to run in a network with several computers, using all the processor for the calculations. This was done to decrease the computational time of the runs and to explore a more extensive analysis with a bigger number of iterations. The experiment was setup using five laptops with the same configuration. The specifications of the machines are presented in table 3.14 and table 3.15 presents the number of test that were run and the total of iterations that were assessed.

Table 3.14 Laptop' s Characteristics

Characteristic	Value
Operative system	Windows 7 Enterprise - 32 bit
Processor	Intel Core i5 2.4 GHz
Number of processors	4
RAM	4.0 GB
Hard Drive	300 GB

Table 3.15 Trial runs conducted for the calibration process

Test #	Number of Generations	Population Size	Total Runs	Decay Function
1	15	336	5040	YES
2	15	336	5040	NO
3	20	224	4480	YES
4	20	224	4480	NO
5	90	224	20160	YES
6	90	224	20160	NO
7	20	1008	20160	YES
8	20	1008	20160	NO
9	30	1008	30240	YES
10	30	1008	30240	NO
11	50	2400	120000	NO
12	1	2400	2400	NO

Table 3.16 shows the number of variables to be optimized through the genetic algorithm for the entire experiment.

Table 3.16 Number of variables in the calibration process

Parameter	Number of Variables	
	Model M1	**Model M2**
Expander		
Mean Patch Size	16	12
Variance Patch Size	16	12
Patcher		
Mean Patch Size	16	12
Variance Patch Size	16	12
Modulate Change Matrix	16	12
TOTAL	**80**	**60**

The results for all the tests are presented in table 3.17 and in the figures 3.34 and 3.35 where it is also included a comparison of improvement against the fitness obtained with the default and the adjusted values.

Table 3.17. Fitness Results from the Optimization Process

Framework	Model M1			Model M2		
	Fitness	Fitness default	% Improvement	Fitness	Fitness default	% Improvement
1	0.3709	0.37	0.78	0.2800	0.276	1.42
2	0.6160	0.58	6.14	0.4650	0.47	-1.66
3	0.3796	0.37	3.15	0.2810	0.276	1.78
4	0.6054	0.58	4.32	0.4730	0.47	0.04
5	0.4250	0.37	15.48	0.3010	0.276	9.02
6	0.6240	0.58	7.52	0.4840	0.47	2.36
7	0.4243	0.37	15.29	0.2850	0.276	3.23
8	0.6236	0.58	7.45	0.4770	0.47	0.88
9	0.4261	0.37	15.78	0.3040	0.276	10.11
10	0.6239	0.58	7.50	0.4840	0.47	2.36
11	----- (1)	-----	-----	0.4950	0.47	4.69
12	0.6040	0.58	4.08	----- (1)	-----	-----

(1) ----= Not Modelled.

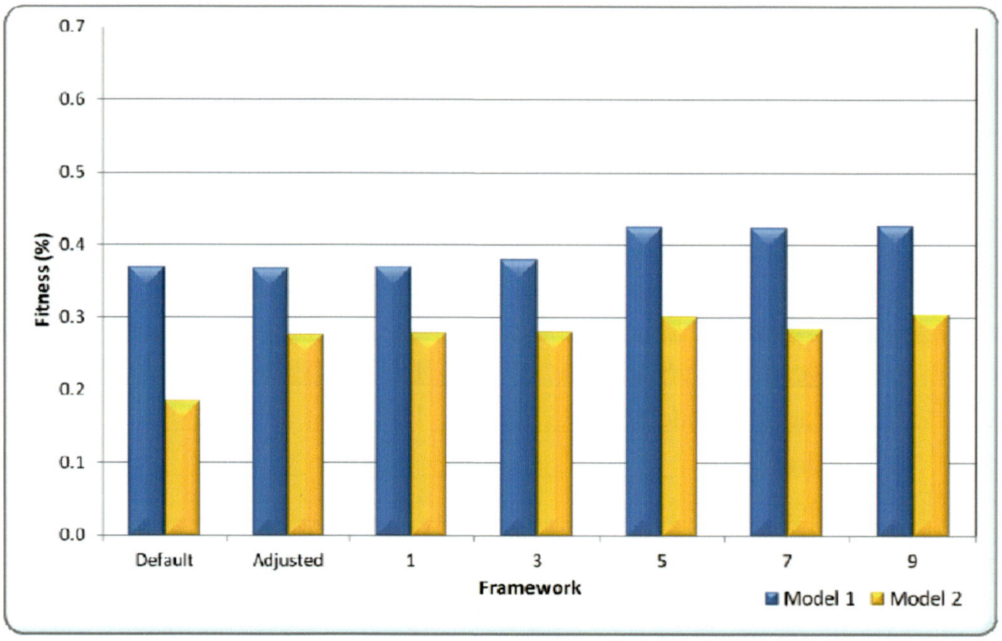

Figure 3.34 Fitness from the optimization Process, with decay Function Models M1 and M2.

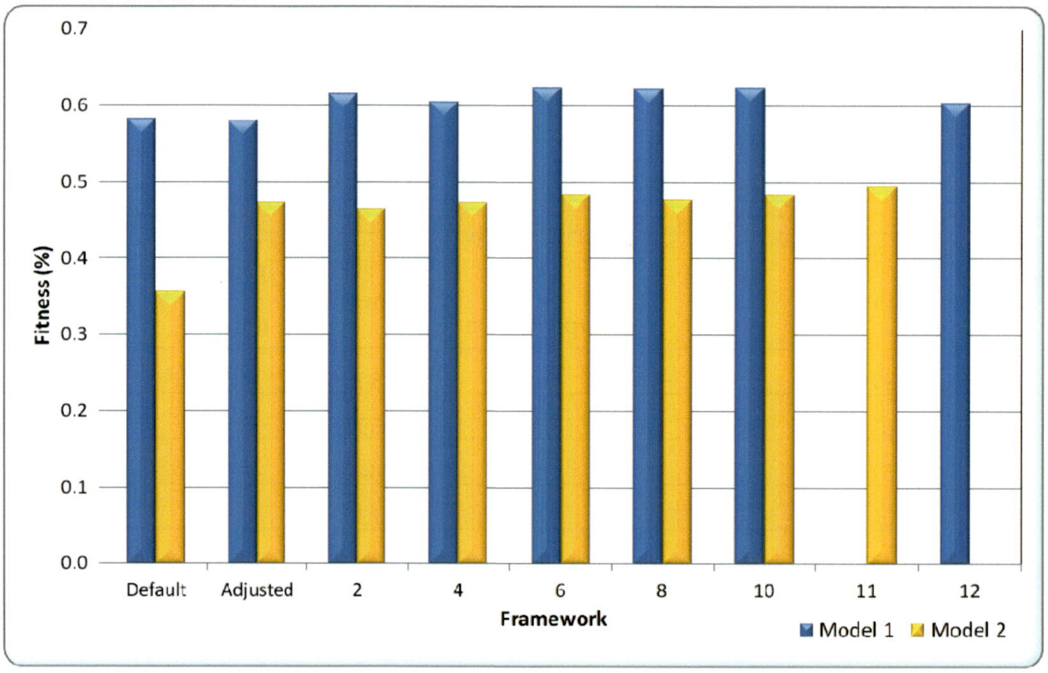

Figure 3.35 Fitness from the optimization process, without decay function models M1 and M2.

The results of the calibration process shows that it was possible to obtain a fitness value of 62 % for model M1 without considering the decay function or 43% considering it. The relative improvement achieved by using the genetic algorithm with respect to the simulation using the default values is around 15%.

In the case of model M2, the similarity or fitness was 48.4% without decay function and 30.4 % using it. For this model the relative improvements are in the range of 31 to 64 % in comparison with the default simulation.

In all the conducted experiments the model M1 always performed better than model M2 no matter how the fitness indicator was computed, with or without decay function. This is possible an indication that the resolution of the model plays an important role in the dynamics of the model.

When the calibration process finished the best parameters were chosen to run again the simulation model. The best sets of parameters were those that produce the highest fitness between the simulated and final landscape for each model. In the case of model M1 the parameters from the calibration test number 10 was chosen and for model M2 the test number 10. The values for all the variables that were calibrated are presented in annex 3. Figure 3.36 presents the final landscape vs. simulated map for model M1, while figure 3.37 presents the same maps for model M2.

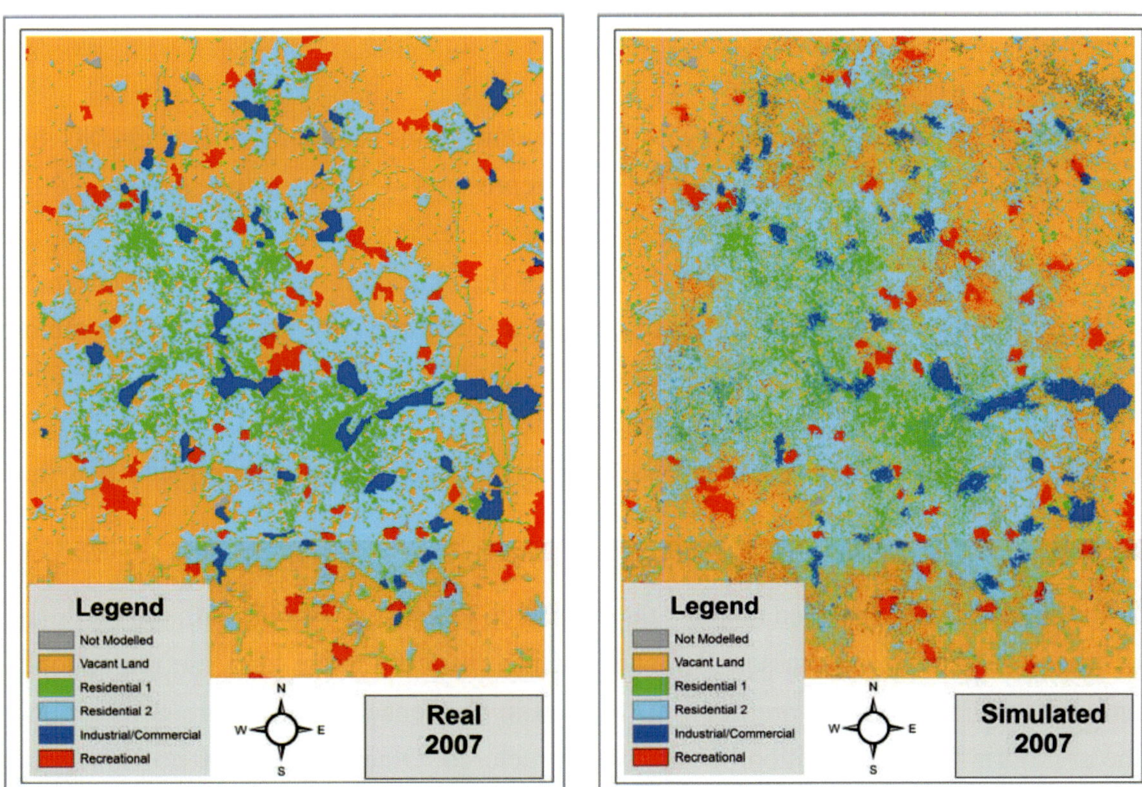

Figure 3.36 Final Landscape vs. Simulated Map from the year 2007, Model M1 (30m).

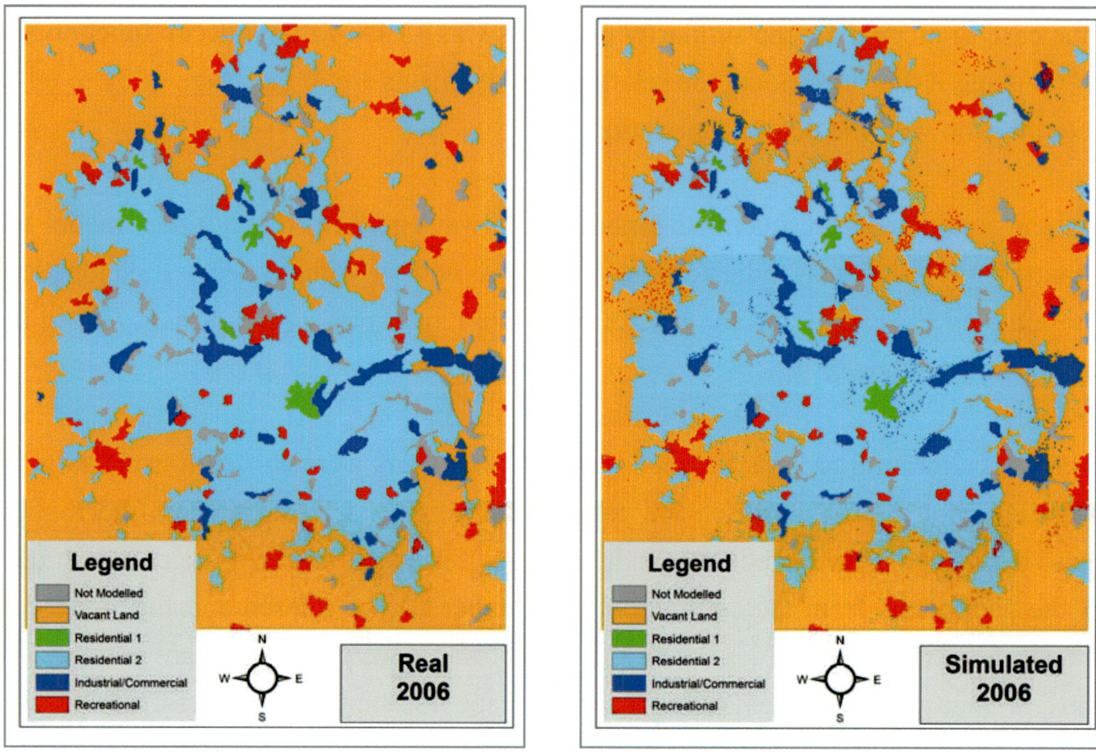

Figure 3.37 Final Landscape vs. Simulated Map from the year 2006, Model M2 (100m).

A visual inspection to figure 3.36 with the outcome of the simulation of model M1 for the year 2007 show that the simulated map represents well the actual map of 2007. There are some differences, for instance for the classes residential 1 it can be observed that the model tends to spread the cells all over the map, perhaps trying to create new clusters without success. Something similar occurs with the class residential 2, it seems like the spatial position is correct but the clusters are a bit diffuse. For the industrial/commercial class some clusters were not well simulated or formed. The clusters of the recreational class are the best formed, but it has to be noticed that this class has the lowest rate of transitions.

For model M2, the simulated map for the year 2006 looks very similar with the real map of the same year for all classes. But some of the clusters for the land use recreational and industrial/commercial are not well formed in the sense that there are cells around the appropriate spatial position but not well structured. Other clusters are not simulated at all. A visual comparison of the simulated land use maps for model M1 and M2 show that the model M2 look better structured, at least visually. In terms of the fitness indicator model M1 is better compared with model M2. This can be explained due to smaller resolution of model M2, in which the information is more aggregated.

Regarding with the numerical fitness of the simulated models, some authors refers as a good index of performance values close to 0.4, representing a good level of compatibility between the real and the simulated scenario. In the literature as well, are reported values of fitness when modelling land use change under the same approach of Dinamica EGO, those simulations were done in less complex simulations in terms of extent, cell size and number of classes to simulate, the reported fitness (Soares et al. 2012) are in the order of 0.4 and 0.9. Adams, 2008 and Ferrari, 2008 who obtained 0.44 and 0.84 for the dynamic simulation models for land use and land cover respectively, and Benedetti, 2010 which reached levels from 0.64 to 0.99 who simulates forest development. These values suggest that even though this research work in a complex scenario in terms of cell resolution and number of classes than traditional uses of the software, the values of similarity or fitness obtained are good. This also suggests that the model can be used for the simulation of future scenarios of change with some confidence.

✓ **Performance of the parallel computing.**

Depending on the problem configuration, number of generation, population size and number of computers used to setup the cluster computer there is an increase in the speed of the calculation, it can be performed from 1.5 times to almost 9 times faster in comparison with a single execution time of the model. Despite the significant increment on the number of evaluations of the land use change model, comparing the initial run of the model for this case and the updated versions of it, the overall fitness performance is the same, in particular for model M2 with a resolution of 100 meters.

3.4.7 Simulation Future Scenario (Year 2040)

Using the best parameters of the calibration, the model was set up to run the future land use simulation, corresponding to the year 2040 in this research. Figure 3.38 and figure 3.39 presents the results of such simulations for model M1 and M2 respectively.

From the simulated maps until year the 2040 in both models, it is shown that under these scenarios, the tendency of growth of the city of Birmingham is to increase the land use Industrial/Commercial in those areas close to the city centre. Recreational areas growth around the actual patches of this class and a clear expansion of Residential 2 to the outsides of the city is evident. Residential 1 remains almost constant.

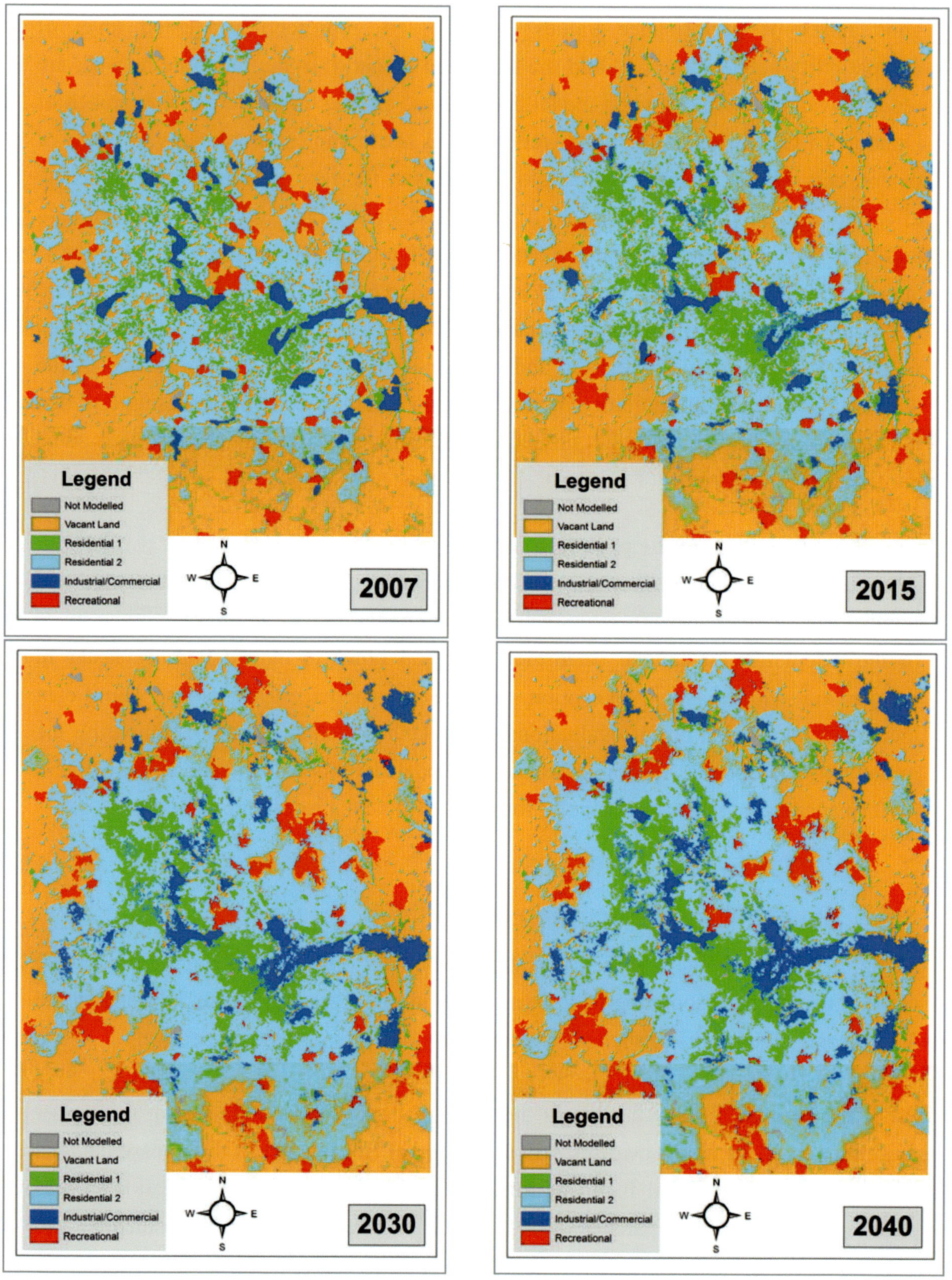

Figure 3.38 Future Land use Maps, Model M1 (30m)

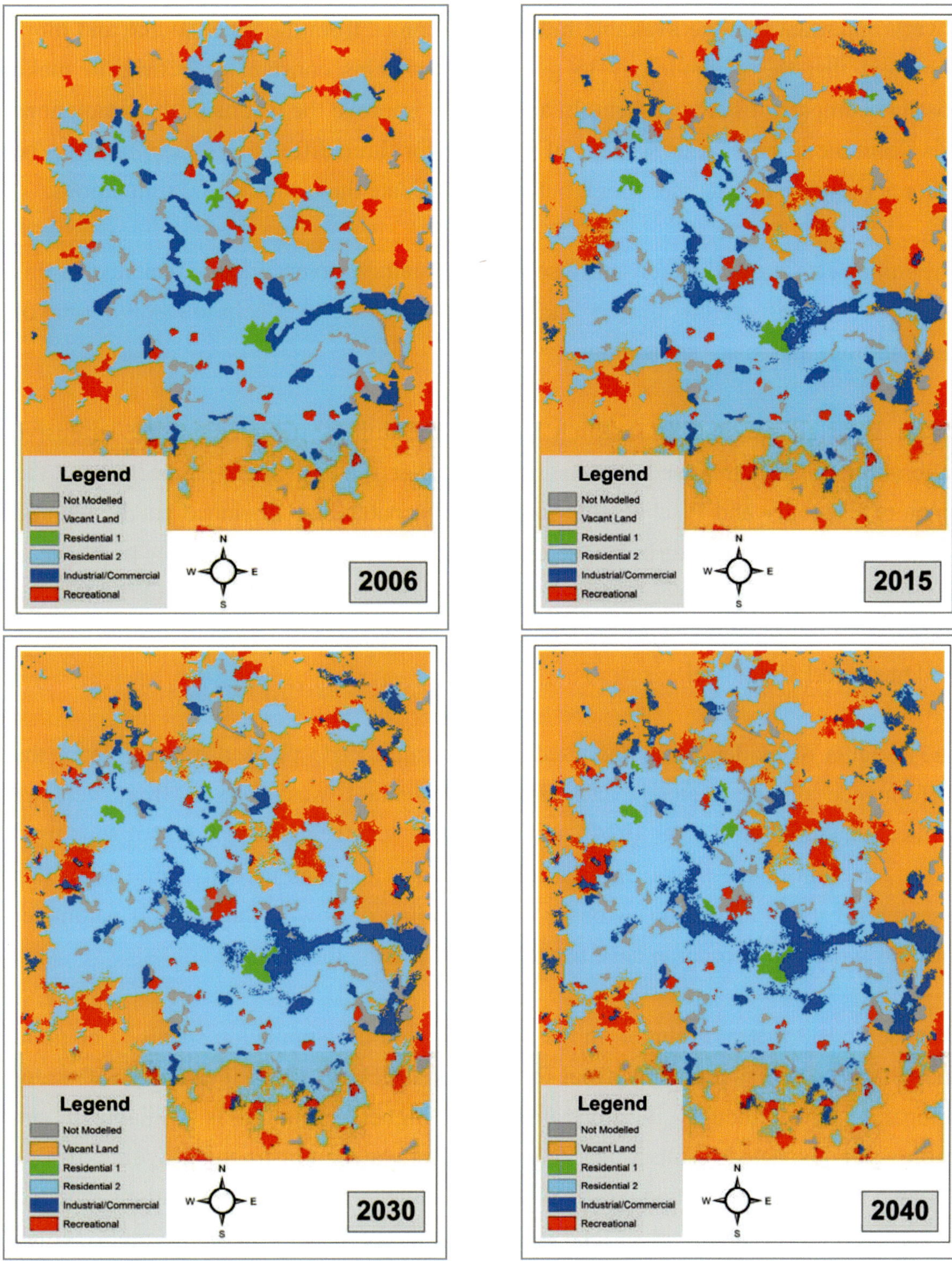

Figure 3.39 Future Land use Maps, Model M2 (100m).

3.5 Conclusion

Dinamica EGO, and cellular automata in general proved to be an excellent tool for gaining understanding and simulating urban land use change. But, as in any model the accuracy of the simulations is directly related to the data used. To have better models it is very important that the available information comes from the same source and more important that the methodology to generate it remains constant over time to make possible comparisons among them. In addition, to improve the results of the simulations it is desirable to have at least maps of land use for three different years in order to be able not only to calibrate the model, but also to validate the results of the simulations.

The initial data collection for land use maps contained severe inconsistencies. In this respect, the use of satellite imagery such as Landsat images can help to resolve some of the omissions and misclassifications.

It has been shown in this research through the fitness obtained in the extensive calibration process followed in this research, that the transition parameters of the functors Expander and Patcher, referring to mean patch size, variance patch size and isometry, as well as the Matrix of Change, only control the simulation of the phenomena of urban expansion up to a certain level. To obtain better fitness in the simulated maps it is necessary to adjust other parameters such as the weights of evidence coefficients in the Dinamica EGO model.

The weights of evidence method used by Dinamica EGO plays a major role in the performance of the method, and since this method was derived from geology applications (Bonham-Carter 1994), and in geology science the formations of soils used in the simulations are bigger in area than the ones that was used to simulate in this research for urban expansion. A better understanding of how to tune up this parameter must be addressed in order to have better results by using this method.

The use of 100 meter resolution in the model of land use simulation seems to be good enough to be used as a planning tool for municipalities, land use developers and water companies. Even though, if information with higher resolution is available it can improve the expected results when deriving urban water infrastructure layouts. In this research a 30 meter resolution was used, which is one of the highest resolutions reported in the literature to model urban land use change at the scale of an entire metropolitan city and its surroundings. This can represent a good basis to improve future land use planning at a more detailed scale.

The implementation of parallel computing in the problem of optimization of land use change simulation, in combination with genetic algorithms (NSGA-II), has proved to be a promising technique that can save a great deal of time and to reach good set of possible solutions.

4 Evolution of water distribution networks

4.1 Introduction

Urban water distribution networks are complex and dynamic systems that consist of components which interact with their physical environment and the social fabric of society. All the possible interactions are not completely understood and can be hard to quantify. In particular, the analysis of water use by a city is often oversimplified by the assumption of demand patterns. While these are simple to use in practice, the estimation of the reliable patterns for a large variety of possibilities is difficult to realize. How water is used in a network depends in part on the behaviour of the community. Culture, the variability of the weather, and the use and adaptation of new technologies play an essential role. The total demand of the network is one of the key components for the design of the network since the size and total cost should be directly linked to the total demand.

Integrated urban water management is challenging because it aims at the sustainable use of water resources so that present demands as well as future demands can be met in terms of quality and quantity. The current practice in the water sector is not optimal. Considerable amounts of water are lost through leakage from water distribution networks. There continues to be unacceptable pollution of receiving water courses from combined sewer networks. Use of energy is inefficient. There is poor interaction between town planners and water engineers concerning urban expansion. These and other issues call for innovative thinking and the adaption of new strategies including integrated planning and development. The idea behind this research is the development of methods and tools that can help planners and decision-makers at the city level understand the main drivers affecting the urban water cycle, analyze future scenarios for the city expansion, foresee bottlenecks and generate possible solutions. The tools can be used for rehabilitating water networks, testing future scenarios and planning sustainable strategies.

A model for land use change is adopted to estimate the future layout of the water distribution network in possible areas of expansion of the city. Urban areas are considered to be complex systems based on their characteristics of emergence, self-similarity, self-organization and the non-linear behaviour of land use changes with time; Batty et al 1997. The use of tools in helping to understand the dynamics of these characteristics is important in order to gain insight into their patterns and mechanisms.

In particular, cellular automata (CA) can be used to model these urbanization phenomena.

4.1 Considerations for the design of water mains

A water distribution system consists of a catchment where a source of water is located, the source (river, aquifer, lake, etc), a water treatment plant, storage tanks, pumping stations, pipes, valves, etc. The water demand depends on the living standards of the population, weather, habits, culture, etc.

Traditionally the planning phase of water distribution network involves consultation with several stakeholders and authorities at municipal and regional level. Socio-economic plans and specific information about the existing water supply systems can be obtained based on these consultations,. Information about the spatial locations of the future expansion of the urban area and the locations of possible water sources can help estimate present and future water demand based on population data, housing, and the industrial and commercial planning. Due to the uncertainty in many of the factors affecting the future development, it is common practice to develop water distribution facilities in stages or by a sequence of master plans. This practice provides opportunities to assess and adapt the design of the possible expansions of the network where they may deviate from original concepts.

The design of water distribution networks requires consideration of hydraulic and engineering criteria. The hydraulic performance is assessed in term of the provision of water to meet the demands of the population, and acceptable pressures and velocities in the pipe network. It also must ensure adequate functioning of the network during emergency events (fire, pipe burst, etc) such that the operational cost is maintained at a low level. The engineering criteria are also important to ensure the durability of components of the system during the life expectancy of the network, including pipe materials, valves and pumps, and the construction material of other components such as tanks or reservoirs.

The hydraulic design requires detailed calculations because the performance of each component affects the operation of the others. The layout of the network is the first step in the design and it directly affects the costs, performance, operation and maintenance of the system. It is common to have looped networks in urban areas, more so than in regional systems. Once the layout is defined the sizing of the system and the different components can be considered as an optimization problem.

The main problems related to water distribution networks are the ageing of the pipes that can lead to increasing leakages and pipe bursts, aggravated by the growing demand for water due to population growth, urbanization and economic development. Climate change is causing a disruption to the availability of water resource from different sources so that demands cannot be met and water has to be transported from further afield.

4.2 Integrated and strategic planning

Strategic plans for urban water distribution networks need to take on a long term perspective (15-40 years) because the life cycle of much of the network is typically 40 years or longer, and because the changes and pressures have become more acute during this period. Some changes occur gradually, but other changes may occur in steps. The planning process needs to take into account the uncertainties in these changes and therefore needs to be built on a flexible strategy, using technologies and methods that are adaptable and that can be applied under different future scenarios.

Projecting and simulating the land use changes in space and time is crucial to understand and assess consequent environmental impacts. The simulation of humanly - influenced landscape changes following different economic, cultural and demographic scenarios is helpful in order to reveal strategy policies that can be modified to improve environmental issues of the future. Nevertheless, planning scenarios for long term policies (beyond 15 years) may not be practical since other processes related to macroeconomics and politics make the scenarios uncertain.

Planners and policy makers face a difficult task since the world they must deal with is complex, dynamic and highly interconnected. This is the case in urban planning where the physical and the human systems are deeply interrelated. Understanding the processes that drive changes in these systems is the key in the formulation of effective management policies.

There are certain aspects that are important to understand in urban planning:

1. The system is integrated. Whereas the decision makers and planners may intervene directly in a limited part of the system, the connections with different parts of the systems will propagate the consequences of such actions for good or ill.
2. Human systems are not in equilibrium: they are changing continuously. Certain small actions can trigger unanticipated changes in the system.

3. The physical and human systems are spatially varying: therefore the consequences of planning policies depend on the spatial contexts within which they are implemented.
4. Urban systems are uncertain, so that, even with a good understanding and knowledge of the system and with improved modelling tools, uncertainty cannot be eliminated.

Scenario planning is a systemic method that has been tested in several fields as a tool to generate scenarios that can deal with complexity and uncertainty. Petterson et al, 2003; Kok and van Delden, 2007.

4.2.1 Scenarios and scenario planning

A scenario describes possible future, futures that could be rather than will be. Peterson et all, 2003. In this sense scenarios describe a vision of possible future pathways considering critical uncertainties. Scenarios can be defined as storylines that can be told in words and numbers which offers a description of how events can evolve over time, like a movie. Kok and Van Delden, 2007. This makes scenarios an easy tool to work with stakeholders and community leaders in participatory process.

A common use of scenario derives from computer models. The configuration of such models depends on assumptions about the extrinsic drivers, parameters, and structure of the model. Variations in the assumptions used to create such models are often described as scenarios. For example, the differences between the climatic scenarios of the Intergovernmental Panel on Climate Change (IPCC) are determined by differences in assumptions about demography and social, economic, technical, and environmental development.

Scenarios should be differentiated from predictions, forecast and projections. While scenarios are something possible; predictions, forecasts or projections are something probable. A prediction is understood to be the best possible estimate of future conditions. Whereas scientists understand that predictions are conditional probabilistic statements, non-scientists often understand them as things that will happen no matter what they do.

In contrast to a prediction, a forecast is the best estimate from a particular method, model, or individual. The public and decision-makers generally understand that a forecast may or may not turn out to be true. For example the weather forecast.

A projection is something which may be heavily dependent on assumptions about drivers and may have unknown, imprecise, or unspecified probabilities. Projections lead to “if this, then that” statements (MacCracken 2001).

Kahn (Kahn and Weiner, 1967) and others were pioneers in developing scenarios for strategic planning applications. Scenarios were refined at Royal Dutch/Shell by Wack in the 1970s and 1980s, and Shell became a leader in the scenario approach to business planning, Peterson et al , 2003; Kok and van Delden 2007.

Today scenario development is used in a variety of different contexts ranging from political decision making, to business planning. Increasingly, decision makers and other stakeholders are being involved in the process in different ways (see Van Asselt and Rijkens-Klomp, 2002). The combination of quantitative models and more qualitative participatory methods to develop scenarios is currently being advocated as a promising way forward (Alcamo, 2001; Kok et al, 2007). According to Wilson (1998) the golden rule in deciding the number of scenarios is no less than two, and no more than four. As well, Wilson (1998) and Wollenberg et al. (2000) explain that the following five criteria could be helpful in generating a scenario:

- Plausibility: the selected scenarios have to be feasible
- Differentiation: they should be structurally different and not variations of the same theme
- Consistency: the combination of logics in a scenario has to ensure creditability
- Decision-Making Utility: each scenario should contribute specific into the future decision making.
- Challenge: the scenarios should challenge the organization’s conventional wisdom about the future.

Advantages

A scenario is a portrait of a possible future world that introduces a variety of new ideas and lead to a gradual understanding of what the new situation means. The advantages of scenarios is also that they do not describe just one future, but that several possible or even desirable futures are placed side by side in order to learn to deal with what might happen and they all should be equaled considered.

The scenario approach can help regarding the improvement of the learning process, the improvement of the decision making process, and the identification of new issues and problems which a system may have to face in the future (Martelli, 2001).

Disadvantages

Without a clear direction the numbers of possible scenarios are almost endless and the key factors and description are difficult to limit. So that we should ask: What planning questions need to be addressed? What variables are most likely to forecast in order to address these concerns? (Mietzner et al., 2004). Also, the practice of defining and assessing scenarios is very time-consuming and resources demanding. Barredo et al, 2005; Kok and Van Delde 2007

4.2.2 Integrated urban water systems modelling

One dilemma that any intervention in urban infrastructure systems, like water distribution, needs to face is the fact that known current condition of the system is based on the past, and the decisions done now are about the future. In this respect, "we live in old cities" in the senses that they were planned 25 years ago. There is a need to develop tools that help to understand the process and drivers of the human "ecosystem" that made up an urban area. Uncertainty in urban planning and the water sector is high and cannot be reliable. One of the main issues is that the required information and data to assess the condition of the system is often non-existing or incomplete. Therefore is not convenient to relay merely in the forecast of factors that influence water supply and water demand. One option is to use scenarios and scenario building as an integral part of the strategic planning process.

Urban water distribution networks are complex and dynamic systems containing several components that interact among themselves and with the physical environment. Many of the interactions are poorly understood, in particular those in the sphere of the socio-economic layer, and hard to measure. more importantly, is that the data that is gathered correspond to conditions of the system now and not how it will behave in the future, nevertheless this data is important to understand the overall system behaviour.

Traditionally the approach for water distribution networks planning and design consists of doing an estimation of the future water demand and the compliance to a set of regulations that are set by the local authorities. The estimation of the future water demand considers the projection of the population growth and the possible amount of water use per capita, which depends itself on a range of factors including economics and technology.

The performance of the system is determined based on the fulfilment of the objectives set for its operation (water demand, storage, continuity, water quality, emergency flows, etc). Each individual component in the water distribution system can be repaired upon failure or replaced at certain cost. The decision to repair or replace a component in the system is made by determining the minimum cost that can sustain or improve the performance of the system. The notion of a single pipe or element performance has no meaning as an isolated component, to evaluate the performance of the component and its life service make senses in the integrity of the system as a whole.

The long term planning for upgrading and rehabilitating the water distribution systems includes the selection of the best combination of alternatives for the different components (Pipes, valves, tanks, etc) and the implementation timing (immediate action, mid-term, long-term, etc) while keeping the level of service. The problem is complex due to the fact that aging and deterioration of the components lead to a diminishing hydraulic capacity while the cost for operation, maintenance and repair will increase over time. Moreover, the future expansion and layout of the system is not known, this is an important issue since the enhancement or change in the capacity of one pipe will cause a redistribution of energy and flows within the network. The performance of a single element also depends on the state of all the other elements in the network. The rehabilitation or upgrade of a component in the systems will have a future influence in which elements need to be rehabilitated, when, and the range of alternatives that can be selected.

4.3 Data Requirements

The application of the developed techniques requires a dataset that involves collection of data for land use/cover from at least two different years in order to determine the changes in different land use classes. Other datasets required include ancillary maps that provide data to explain the process of urbanization or land use change, such as elevation, slope, boundaries, road network, or other services including gas, electricity and cable television, rivers, water bodies, etc. Depending on the availability of maps showing plans for future housing developments, business and industrial investments, and road expansion, these can be used to assess future scenarios.

Some of the constraints involved in using this modelling paradigm come from the limit on the amount of data required and the type of data. The implementation of satellite and remote sensing techniques has enhanced access to spatial information globally in recent years. Some issues still remain, for example the type of sensors used and the objective of the project are dependent on the time the information was capture and therefore influence the quality of the dataset. Also the classification of land cover/use for agriculture purposes differs from that used for urbanization. In both cases the built-up area can be identified, but the land use within the built-up area requires ground-truth verification points for instance commercial and industrial areas.

Other issues are related to the fact that by using remote sensors the available datasets refer entirely to the physical world. In other words, is possible to have information about the environment, boundaries and some infrastructure but it is not possible to map the decisions taken by political actors, decision makers, institutions, the economists, the business market, etc. All of which can be important in setting the rules that drive urbanization. The work presented here makes use of information and datasets that are currently freely available on the Internet together with other datasets that are acquired specifically for the case study.

For example, in the case of urban areas in Europe land use maps can be obtained from the European Environmental Agency corresponding to the Corine Dataset. The dataset exists in different resolution for the years 1990, 2000 and 2006. Since the focus of the study is on urban areas a higher level of detail is considered necessary for artificial areas and therefore the Corine land use classes were regrouped into 6 classes mainly bringing together the agriculture areas, forests and pastures into a class referred to as 'vacant land'. The other classes inside the urban agglomeration are: the continuous urban fabric which was renamed 'residential 1', the discontinuous urban fabric was renamed as 'residential 2'; and the industrial/commercial fabric was kept the same together with recreational land use. Airports, water bodies and construction sites were separate Corine land use classes and were regrouped into a single class called 'not modelled'. They were considered static classes in this study, that is, they did not change during the analysis.

Two different sources were used for the terrain data: Shuttle Radar Topography Mission (SRTM) data with 100 meter resolution and ASTER Global Digital Elevation Model (ASTER GDEM) data with 30 meter resolution. Both DTMs were projected and used to generate the slope raster maps for 100 and 30 meter resolution.

Other datasets representing the motorways, trunks, primary and secondary roads, rails, rivers and canals were downloaded from the open street map project (http://www.openstreetmap.org/). Other sources of information are the municipality's office or local government. Maps deploying land use and basic infrastructure can be acquired through their websites.

4.4 Relations between water distribution networks and land use

One of the main questions in water distribution planning and design is how to extend the system to supply the demand of new urbanized areas so that the impact to the existing system is minimum?. The extension of the system requires to be analysed carefully since a massive intervention in the existing system can disrupt many other services and networks that the cost needs to be incorporated in the analysis.

To develop a connection between the land uses and the existing water system a corridor analysis was done along the main pipes, see figure. This process consisted of selecting the pipes with a diameter bigger than 356 mm and creating buffers along those pipes every 100 meters for a total of one kilometer; this was done in ArcGis9.3. The layer with the buffers was then intercepted with the land use map for the year 1991. This map consisted of 11 classes, including: undeveloped land, residential, commercial, industrial, institutional and recreational categories, among others.

The information can be used in particular to predict the positioning of new pipes based on the estimated land use. The land uses Industrial and institutional are more sensitive to the distance from the main pipes and therefore the distribution of this land use category can be used in determine the layout on the trunk mains both existing and future. These results can be used to position new pipes in future developments by knowing the predicted spatial distribution of land uses, in particular, commercial and industrial areas. This also has the implication that the land use change model must be capable of simulating the internal dynamics in the city. This capacity has been proven in several cases by different studies as described in the previous section.

The links between the spatial distribution of a drainage and other features or characteristics of an urban area has been previously studied by some researcher for example Mair et al. (2012) studied the similarities between roads, water distribution and sewer networks. In this study, the detail water distribution network layout was available for three case studies. The information for the road networks was obtained from the open street map project website. The author's found that by intercepting the road network with the water distribution network an average of 81% of the water distribution network length is found within 78% of all the roads. The author's concluded that water distribution network can be re-generated from the road network.

4.5 Algorithms to deduce the route of the water main

The research focus on the evolution and performance of water services networks to evaluate proposed future scenarios that can lead to a better understanding of the actions that need to be taken to achieve sustainability. The hypothesis is that the incorporation of variables related to the water services has an impact in the direction of growth for new developments in the urban areas predicted by a land use change model (CA). These results can then be compared to simulations where no water related services are considered. The results can also be used to describe the impact of the provision of water services as a trigger to stimulate urban development.

The framework for the extension of the water distribution network consists of doing a spatial analysis to find characteristics that connect the land use pattern distribution of an urban area with the properties of the distribution system. In this way, the output of the cellular automata model can be used together with a rule-based algorithm to explore scenarios of future growth and urban dynamics where water is at the center.

4.5.1 Algorithm 1.

A tool that can look to the closest cells with new commercial land use developments can assess the direction to draw the layout of the future water distribution system. This approach is based on the relative distances between the 3 closest points to a boundary point (or connection point selected by the user). The relative position of the points will determine in which direction there will be a main pipe or a branch.

The approach consisted of the following steps:

1. Find the points of interest (Commercial and/or industrial land classes), Count the points, get the spatial position in the matrix (x,y).
2. The Points are shown to the user on the screen. The user needs to input one initial point to start the analysis i.e. the position of the main storage tank.
3. Starting with the boundary point the distance to all the other points is computed. The distance values are organized in an array in descending order.
4. The closest three points to the boundary points are selected.
5. The distance among all the points is computed.
6. A set of nested if...then statements are used to decide the direction to draw a pipe. If the distance between two points is less than 3 (3 accounts for 3 cells of distance) then a midpoint is selected to connect with the boundary point. The

two points are kept out of the array for further analysis. The latest point becomes the boundary point and this is repeated until there are not more points to analyse.

Figure 4.1 shows the results obtained using this approach.

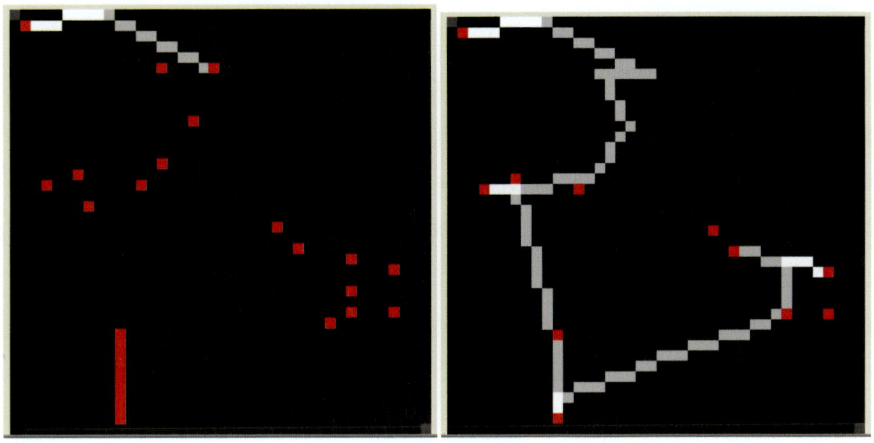

Figure 4.1 Four points approach to extend the layout of the network

The application of the above described procedure works well whenever the points of interest are sparse in the modelling area; but, urban growth and expansion do not occur in house by house building. Often a group of houses is build all at the same time; therefore the above described approach was adjusted to be capable of identifying the clusters of new developments for the particular land use class of interest.

4.5.2 Algorithm 2

Based on the results obtained with the algorithm 1 it was possible to identify that using all the possible attracting cells or points in an urban area within the proposed approach will not work. One of the encountered problems was that the points were very close to each other and the algorithm started to loop between the points and the derived layout was not satisfactory.

A second approach was developed and consisted in the following steps

1 The land use map of the area of interest is read. The user defines the number of clusters to classify the points of interest (Commercial and/or industrial land classes), A version of K-means clustering algorithm was implemented, count the points, get the spatial position in the matrix (x,y).

2 The Points are shown to the user on the screen. The user needs to input one initial point to start the analysis i.e. the position of the main storage tank.
3 Starting with the boundary point the distance to all the other points is computed. The distance values are organized in an array in descending order. Based on this the nearest cluster of points is selected.
4 The closest three points to the boundary points in the initial cluster are selected.
5 The distance among all the points is computed.
6 A set of nested "if...then" statements are used to decide the direction to draw a line. If the distance between two points is less than 3 (3 accounts for 3 cells of distance) then a midpoint is selected to connect with the boundary point. The two or three points "used" are discarded of the array for further analysis. The latest point becomes the boundary point and the steps 4-6 are repeated until there are not more points to analyse in the cluster. A special routine was written to handle the conditions for the last 3 and 2 points of the analysis. Steps 4-6 are repeated until there are no more clusters to analyze.

This approach seems to reproduce a better pattern of the distribution network. Figure 4.2 shows the layout pattern of the network produced by this approach.

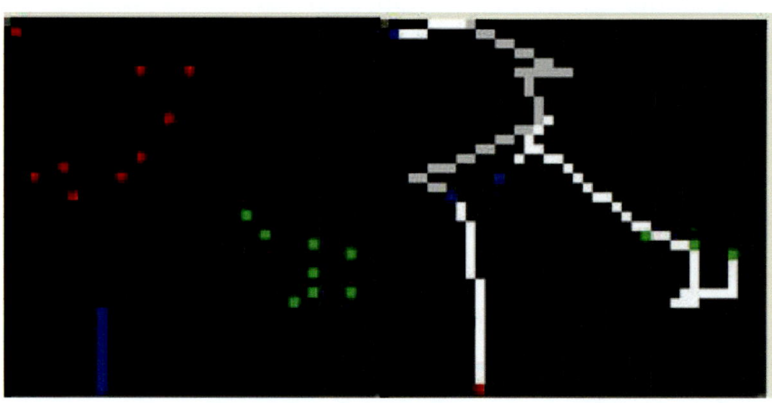

Figure 4.2 Reverse engineering model approach to new network layout

4.6 Sizing and costing of water distribution networks

The cost of any water distribution system is comprised of all direct, indirect and social costs that are associated with:

- Capital investment in system design, installation and upgrading.
- System operation – energy cost, materials, labour, monitoring, etc.
- System maintenance – inspection, breakage repair, rehabilitation, etc.

Water supply system has multiple functional objectives. The assessment of all this objectives are difficult to quantify (e.g., reliability, flexibility, water quality) or difficult to evaluate in monetary terms (e.g., users satisfaction, level of service, sustainability of the system, etc). Perhaps the only exception is the economic efficiency of the system. There are methods to quantify direct and indirect cost in terms of investment and operation and maintenance.

The sizing and costing of the system is formulated as an optimization problem since there is a trade off between the level of investment and the performance of the system.

Since all the above mentioned items will change from location to location an overall average cost of pipe upgrading or replacement per each diameter will be assumed for this study (Pipe Catalogue).

The actual cost of the pipe network Cost 1 is calculated based on the cost per unit length associated with the diameter and the length of the pipe:

$$C1 = \sum_{i=1}^{n} C(D_i) * L_i$$

Where, n is the number of pipes in the network and $C(D_i)$ is the cost per unit length of the pipe (i) with diameter D_i and length L_i.

In an upgrading or design scenario, the objective function to be minimized by the optimization algorithms is the cost of the pipes. If the actual cost of the pipe network is the only objective function, then obviously the search will end up with the minimum possible diameters in the market allocated to each of the pipes in the network. The same logic applies for the maximum value of the function.

In reality, many projects consider the total infrastructural cost to be minimized in the optimization process. The total cost is a combination of actual cost and operational cost when the whole design cost need to be considered. For example cost of storage and water treatment and operation of pumps.

The second objective to be assessed is the performance of the system in this case the supply pressure head boundaries (i.e., minimum and maximum residual pressure head) is used. The total number of nodes where the pressure is above 15 meters during the peak hour is counted and compared against the total number of nodes in the system.

$$C2 = \sum_{i=1}^{n} Node_i (\Pr essure > 15m)$$

The multi-objective analysis is done by using the genetic algorithm NSGA-II (Non-Dominated Sorting Genetic Algorithm) developed by Deb et al (2000). This algorithm has been reported to perform well, solve complex optimization problems, and find good Pareto sets. Since the value of the objective functions can have a wide range and even different units of measures, it is preferred to normalize the functions in such a way that the evaluated values are all in the interval [0, 1]. This will also present an advantage for the graphical evaluation of the Pareto and the solutions. The NSGA II has been couple with the EPANET hydraulic engine to solve water distribution network equations.

4.7 Interface to generate the layout of the system

A graphical user interface was developed in Delphi code gear. All the computational routines were included in this interface, so that the user can better interact with the generation of the water distribution network layout, the preparation of the input file for EPANET and control the optimization process. Figure 4.3 presents a printout of the developed graphical user interface.

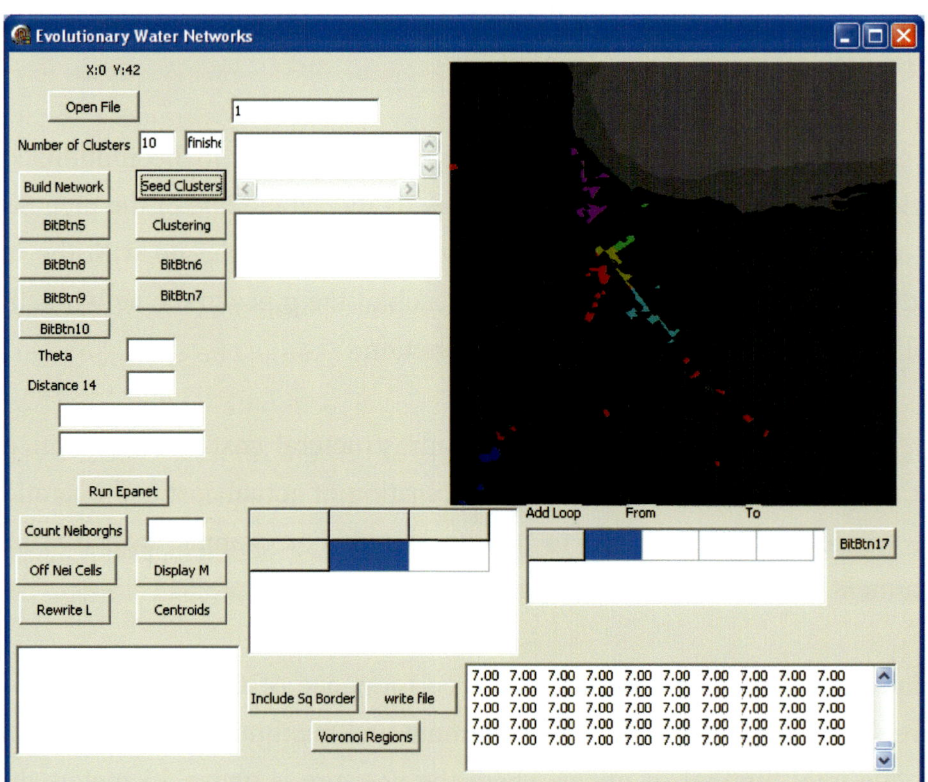

Figure 4.3 Graphical interface water distribution network layout generator.

The functionalities of the graphical user interface include the following:

- Import of the land use map in ASCII raster format, store the information of the land use classes in a matrix and display of the land use map.
- The user needs to specify the number of clusters to be identify in the land use maps according with the land use of interest for a particular case. For example: industrial or commercial. With the bottom clustering a version of the K-Means algorithm is run to perform the classification.
- The bottom "centroids" is used to calculate the centroid of each previously classified cluster of land use.
- The user needs to specify the source for the water distribution network. To do this it is necessary to click with the mouse on the land use map. One point will be generated, this will act as the starting point to build the layout of network and it will correspond to the storage tank further in the procedure when the input file for EPANET is written.
- The bottom "build network" will applied the algorithm 2 described in the previous section to derive the network layout. It will used the starting point specified in the previous step and all the centroids computed earlier.
- Once the layout is finished the "voronoi regions" bottom can be used to estimate the voronoi areas around each node. This procedure is used to allocate and accumulated the water demand on each node. The water demand is associated with each land use class.
- The graphical interface has the possibility to open a new form where the user can add different maps that may be useful in helping constructing the model. for instance the elevation map. With the bottom "transfer values" it is possible to extract the values of the ground elevation for each of the nodes.
- To control that this functionality is working properly, the data is displayed in one of the tables of the main window. The table will display the x, y coordinate of each node, the elevation and the accumulated water demand. Figure 4.4 shows an example of the second form displaying the DTM and the computed voronoi regions.

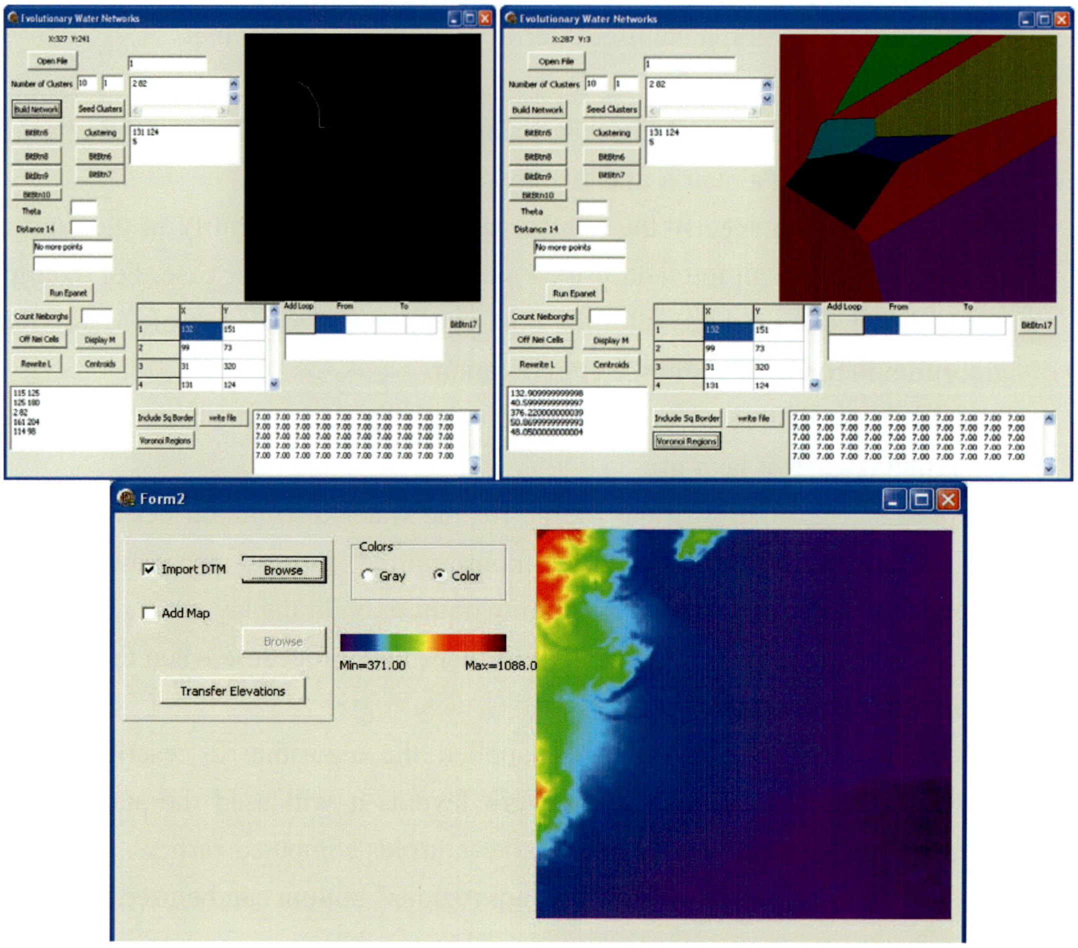

Figure 4.4 Graphical interface with layout, voronoi regions and DTM.

- The bottom "write file" can be used to write the input file of epanet *.inp. The internal procedures will write the required information for each section of the input file. The length of the pipes is computed internally. The main assumptions are that the starting point will act as the source or reservoir. The elevation of this point plus 50 meters head is added to account for the dynamic water losses in the system. This assumption can be modified by the user once the *.inp file is built. A demand pattern is also assumed to build up the model and all the pipes are assigned the smallest diameter available in the pipe catalogue that must be provided by the user in order to use the optimization tool to size the system.
- The "run epanet" bottom can be used to launch the optimization tool. This will open up a new interface where all the parameters to perform the optimization can be specified. Once the optimization is finished the user must select from the pareto set the best solution. For design purposes is preferable to select the one that complies with the minimum pressure in all the nodes. Figure 4.5 shows an example of the optimization interface.

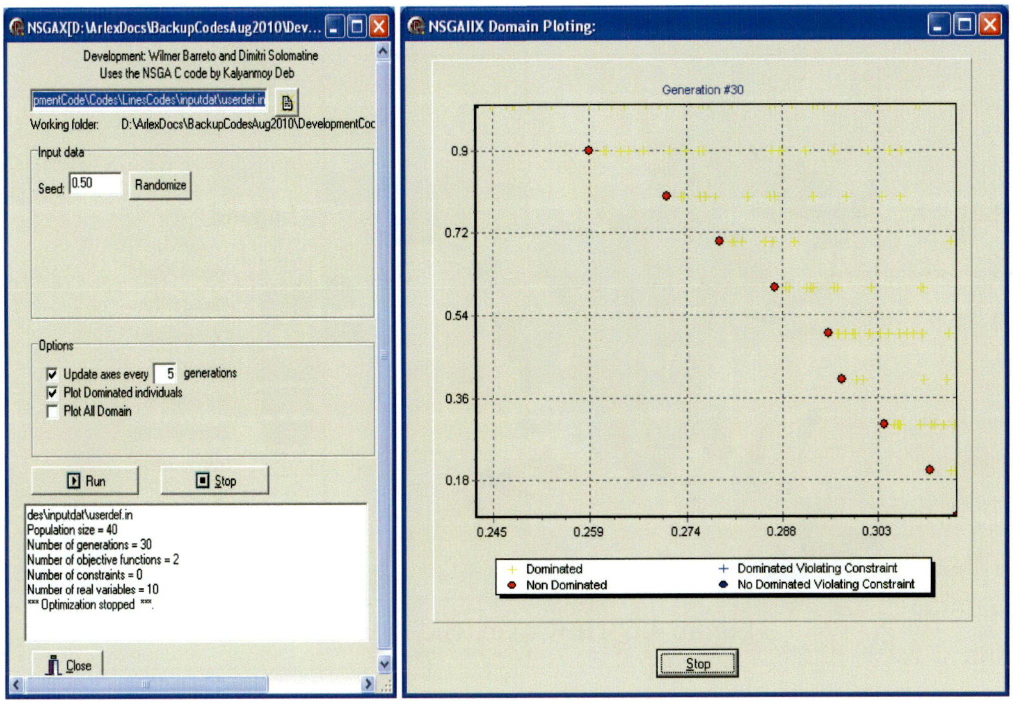

Figure 4.5 Interface for the optimization of the water distribution network

4.8 Case Study 1. Villavicencio Colombia

An experiment based on a case study was carried out in order to test the approach described above. The study area is the city of Villavicencio a medium size municipality located in the south-east part of Colombia with an approximate population of 400.000 inhabitants. Data sets of the land use classification were available for different years. Digital terrain model, main roads and a model for the water distribution system of the city were also available. The basic cellular automata model was built using DINAMICA EGO, a tool developed by the remote sensing centre of the University of Minas Gerais in Brazil to estimate land use changes. The water distribution model was built in Epanet 2.0 and consisted of 4100 pipes and 2800 nodes. The pre-processing of the data set to build the different required maps was done with ArcGis 9.3.

4.8.1 Relation between land use and the water distribution system

One of the main questions in water distribution planning and design is how to extend the system to supply the demand of new urbanized areas in such a way that the impact on the existing system is minimum. Figure 4.6 shows a typical situation were new developments have occurred and the water distribution system needs to be extended.

The extension of the system requires to be analysed carefully since a massive intervention in the existing system can disrupt many other services and networks that the cost needs to be incorporated in the analysis.

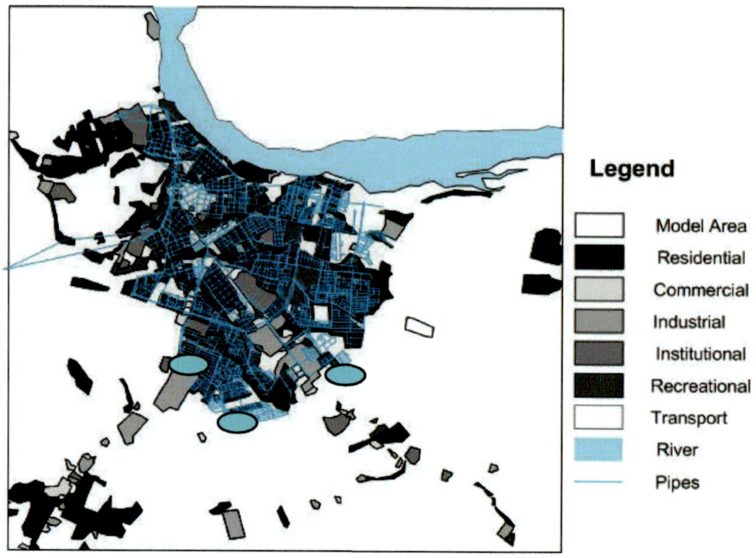

Figure 4.6 How to extend the system.

To develop a connection between the land uses and the existing water system a corridor analysis was done along the main pipes, see figure. This process consisted of selecting the pipes with a diameter bigger than 356 mm and creating buffers along those pipes every 100 meters for a total of one kilometer; this was done in ArcGis9.3. The layer with the buffers was then intercepted with the land use map for the year 1991. This map consisted of 11 classes, including: undeveloped land, residential, commercial, industrial, institutional and recreational categories, among others, see figure 4.7.

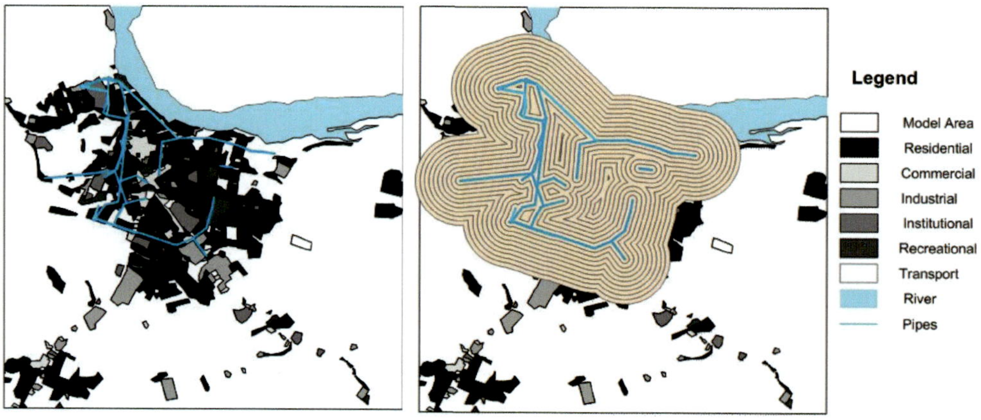

Figure 4.7. Corridor along the main trunk pipes.

Figure 4.8-A shows the result of the analysis for residential land use category. This shows that there is very little variation of the intercepted area of residential land use with the distance from the main pipes. This means that the distribution of residential land use tell us very little about the possible position of water distribution pipes. Figure 4.8-B shows the results for the land uses commercial, industrial, institutional and recreational. This figure shows that commercial and recreational land use can give us significant information about the layout or positioning of the water distribution pipes.

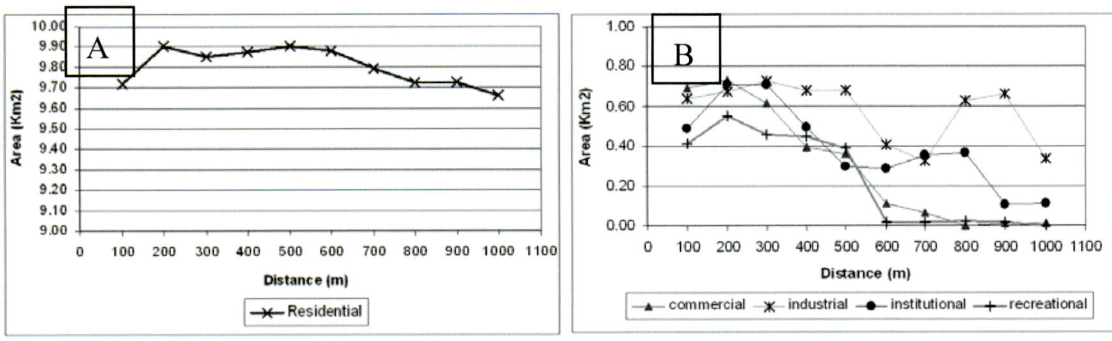

Figure 4.8 A. Distribution of residential land use along the buffers. B. Land use area distribution with distance from the main pipes distribution.

The same analysis using the same pipes and corridors was done for different cell sizes and the results are similar for residential land use no matter the scale of the cells. For commercial the behaviour is also similar. See figure 4.9. The results are more sensitive for institutional, recreational and industrial categories, especially the initial values of the areas covered within the first corridor of 100 meters. This analysis was also performed for different selections of pipes increasing the interval of the pipe size filter to incorporate more pipes and the results were similar.

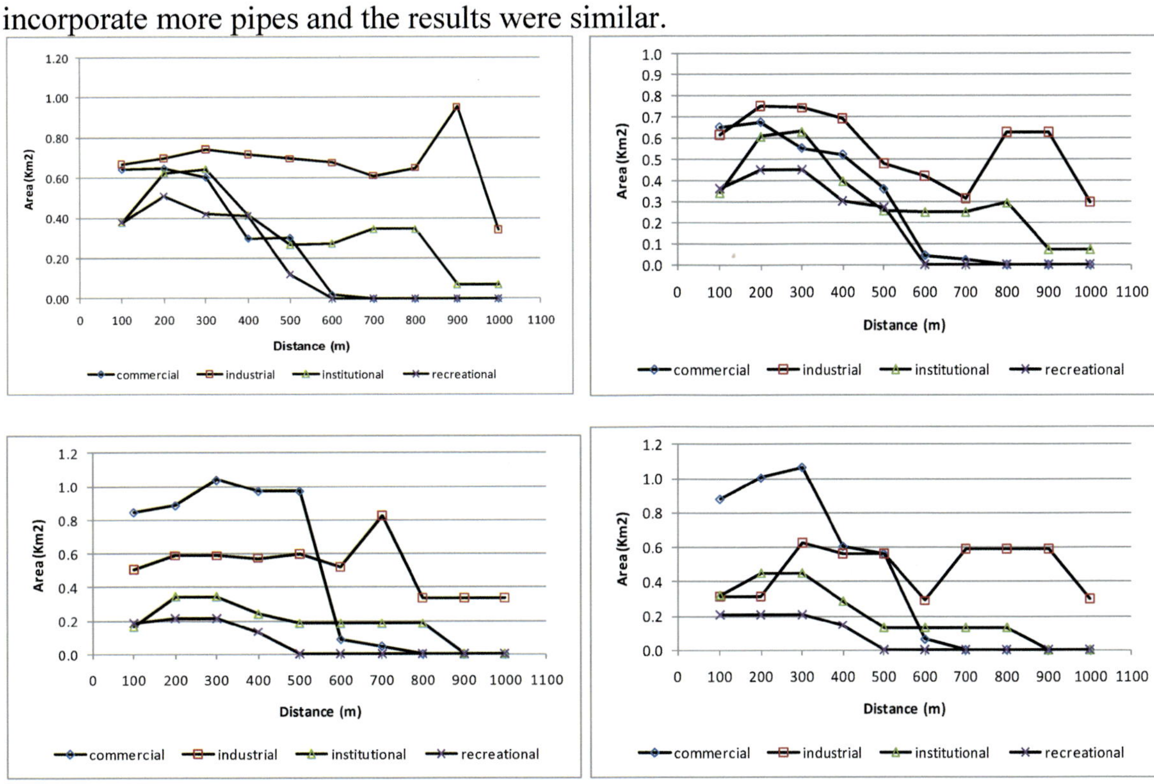

Figure 4.9 Distribution of residential land use along the buffers. The cell size for each set is from left to right and up to down: 100 m., 150 m., 200 m., 250m.

The information can be used in particular to predict the positioning of new pipes based on the estimated land use. The land uses Industrial and institutional are more sensitive to the distance from the main pipes and therefore the distribution of this land use category can be used in determine the layout on the trunk mains both existing and future. These results can be used to position new pipes in future developments by knowing the predicted spatial distribution of land uses, in particular, commercial and industrial areas. This also has the implication that the land use change model must be capable of simulating the internal dynamics in the city. This capacity has been proven in several cases by different studies as described in the previous section.

4.8.2 Generation of the Layout of the water distribution network for the present condition

To evaluate the refined method, the existing system or condition was used as a test. The land use map for the year 1991 was used as a starting point, the algorithm was run to identify 20 clusters or areas with the land use class "commercial". The centroids of the 20 clusters were then computed and used to derive the layout of the network. The starting boundary point that is input by the user was chosen based on the proximity of the existing water treatment plant and storage tank. The idea was then to evaluate how closely this approach can resemble the existing system. Figure 4.10 shows the results.

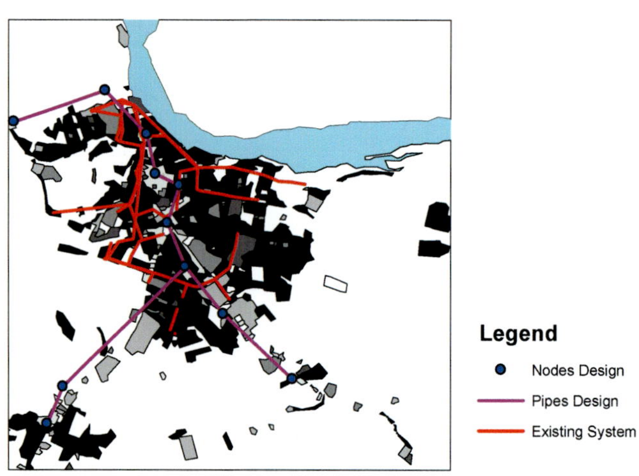

Figure 4.10 Approach to derive the main trunks of water distribution network layout.

The same approach was repeated to the same land use map of 1991, but this time information from the road network was also incorporated. This test was run considering additional points coming from the main roads, with the assumption that the trunk mains follow the main roads. Figure 4.11 shows the result of the derived network versus the existing water distribution system.

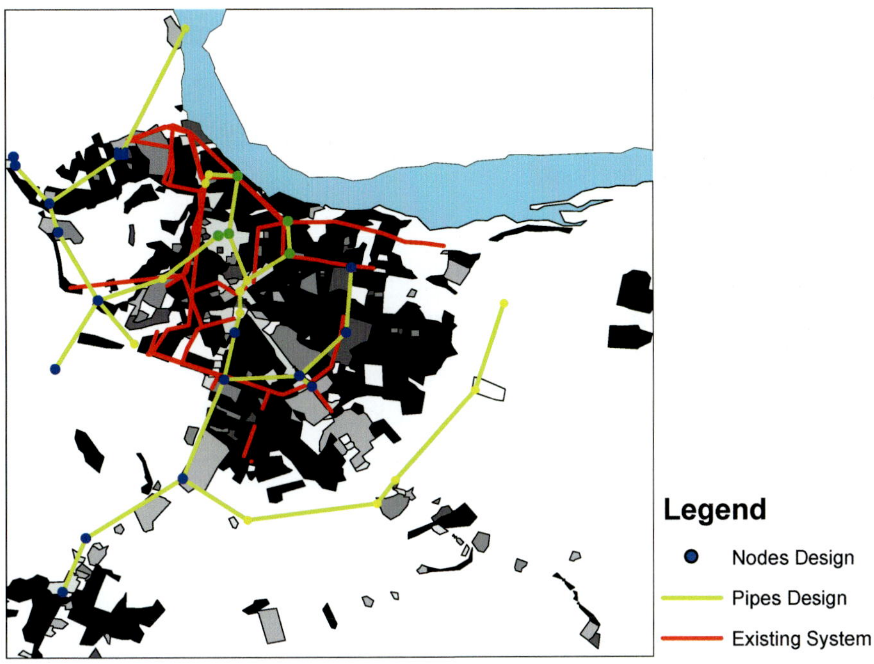

Figure 4.11 Derived water distribution network layout including road information.

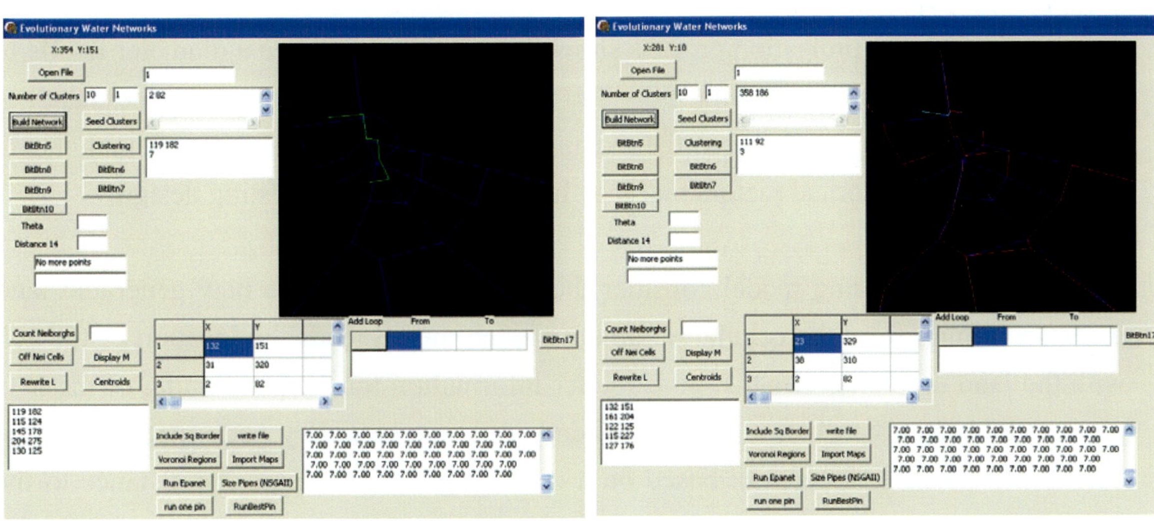

Figure 4.12. Approach to derive the main trunks of water distribution network layout. Left. Using land use information. Right. Using land use and main road nodes.

With the availability of new information is being the model can be improve and validated. A tool that is capable of reading the land use map, for instance the outcome of the cellular automata model, was built. It reads the location of the cell with the land use categories that gives better information for the positioning of the new pipes of the water distribution system; in this case study the commercial clusters were identified. Then beginning with a boundary node specified by the user it generates the layout of the network based on a clustering algorithm for the cells with specified land use categories and the distance to the nearest 3 cluster centers from the latest network node. This process iterates until there are not more clusters left to be integrated. The main idea is to create a layout of the network that approximates the profile found for the different land uses using the corridor analysis.

To test the approach the land use map of 1991 was used as input. The initial point was selected close to the existing water treatment plant. The idea was then to evaluate how closely it can resemble the existing system. Figure 4.12 shows the results of two approaches, one using only the information coming from the land use map and the second considers additional points coming from the main roads, with the assumption that the trunk mains follow the main roads.

The network layout was then used to prepare the input file in Epanet 2.0. To size the system the problem was posed as an optimization problem. Two objectives functions were used, including the cost of the pipes and the number of nodes with a pressure below 15 m.

The results of the optimization process show that all the pipes have a diameter that is in the range of the existing system (350 - 800 mm). There is not a pipe to pipe match in the diameters of the network, but the existing system is not necessarily optimal. There are a lot of other historical factors that may have influenced the existing design.

The reverse engineering model consider the information from the new generated land use map coming from the cellular automata model, then it reads the location of the cell with the land use categories that gives better information for the positioning of the new pipes of the water distribution system. Then based on a boundary node specified by the user it will generate a network based on a clustering algorithm and the distance to the nearest cells and rules to create branches. The main idea is to create a layout of the network that keeps the profile that was found for the different land uses using the corridor analysis that was presented in figure 4.8 B. A first approach for the reverse engineering has been code using Delphi and it's shown in figure 4.12.

This figure shows the application of the clustering approach to the entire map of the case study. The centroids of the clusters were used to derive the network layout. The approach looks promising since it can resemble the original layout of the existing system. Next steps includes the inclusion of other land uses information and sizing the pipes to establish some indicators of fitness of the approach and apply the approach inside the whole modelling framework using different scenarios of futures.

4.9 Conclusion

This study presents the preliminary results of on-going research to develop an integrated approach for the analysis of urban water systems that can be used to design the water distribution services in areas of future expansion of a city. Traditionally this type of analysis is done using different tools in a fragmented way. The new analysis considers corridors along the main trunk pipes of the water distribution networks and the associated distributions of land use to find rules that govern the location of the water mains for new areas of development. An approach that combines information on land use and the road network looks promising; nevertheless the algorithm can extend the layout to areas not covered by the system. Further refinements are needed to improve the performance of the method.

The approach presented here can be used to design the route of the water mains in expected new areas of development. Once the new system is connected to the existing system, the approach can be used to find rehabilitation measures that are required in the existing distribution network to maintain the level of service into the future. Although the work presented here has its limitations; the approach and development of the algorithms show considerable promise, and will be refined and developed further in future research projects.

5 Evolution of drainage networks

5.1 Introduction

Cities are complex systems in terms of their characteristics of emergence, self-similarity, and self-organization, and the non-linear behaviour of land use changes with time (Batty et al., 1994). Several techniques have been used to try and understand the of city systems and infer patterns and mechanisms behind their dynamics. For example, Benenson (1998) used a combined approach of multi-agent systems and cellular automata to study population dynamics in a city; White and Engelen (1994) studied urban dynamics as a self-organizing system using cellular automata (CA). More recently, techniques developed by other researchers based on CA has shown promising results, as such, the technique has been used to model different aspects of land use changes, including urban dynamics (Almeida et al., 2005; Engelen et al., 2007; Liu, 2009; Soares-Filho et al., 2011). From the users' perspective, this technique has several appealing features. The input data can be easily prepared in a GIS environment using raster processing functions. The CA model can be coupled with GIS and the possibilities range from: isolated to, loose coupling to, tight coupling, and to full integration. An integrated GIS-CA environment offers quick prototype creation and an attractive means of demonstration. All the results obtained from CA models can be easily presented and explored with GIS.

An approach to modelling cities has been described by Lechner et al. (2003). The so called Cities Model is a multi-agent based-model which was developed in the framework of the project 'Procedural modeling of cities' at the Center for Connected Learning (CCL) and Computer Based Modeling (http://ccl.northwestern.edu/cities/). This model allows a user to create a terrain and environment in which a set of agents or 'builder' are used to create a city. There are separate agents, each functioning as a builder for a particular type of development, such as residential, commercial, industrial, or park. These agents move through the terrain, grouping patches of land into parcels they are responsible for, or increasing the size of the current buildings. Road builders move through the terrain building roads between the areas, thus increasing the value of the areas and their buildings; (see for example Lechner et al., 2004). This agent-based model was developed in the Netlogo environment (Wilensky 1999). By linking the model with the game SimCity the authors were also able to generate artistic representations of urban landscapes.

Due to data limitations some researchers have proposed the use of virtual or hypothetical case studies to evaluate different approaches or models. For example, Ghosh et al. (2006) propose a methodology to generate drainage networks artificially based on a dendritic Tokunaga fractal tree algorithm in ArcGIS (ANGel - Artificial Network Generator). Möderl et al. (2009) present a tool to generate an unlimited number of virtual sewer systems based on the Galton-Watson branching process. Sitzenfrei et al. (2010) describe a software tool (VIBe) that can be used to generate virtual urban water systems. Initially, VIBe generates a virtual environment including a digital elevation model, roads, buildings, water bodies and so on, and then writes input files for EPANET and EPA SWMM modelling systems to construct models for the water infrastructure.

To date, several case authors have reported successful applications of CA models to predict land use changes at catchment scale (Barredo et al., 2003; Van Delden et al., 2007; Kok et al., 2007). Their results have shown that the application of CA in modelling land use changes is feasible and that the outcomes of the models are similar to what is observed in reality. The Moland framework is an example of such a model that has been successfully applied and calibrated to real case studies in Europe (Barredo et al., 2003). Several other examples of cellular automata models such as Dinamica EGO, SLEUTH and others have also been applied to different cities around the world, as presented by Clarke et al., 1997; Silva and Clarke, 2002; Engelen et al., 2007; Liu, 2009; Zhang et al., 2011; and Soares-Filho et al., 2011, among others.

This chapter presents another new approach to derive the urban drainage network layout by connecting information on land use, topography and other urban infrastructure (e.g. roads). Once the key information from the existing urban system is extracted, the approach combines the use of a cellular automata model of urban growth with an evolutionary optimization algorithm to explore plausible future scenarios and to derive optimal drainage networks. Since the approach can be used to estimate the future drainage layouts by adopting a future land use map, the new system can be sized and connected to the existing system to evaluate the impact of future expansion on the aging existing infrastructure. This approach is tested on a case study in Birmingham (UK). The aim of the present research is to develop tools and methodologies that can be used by planners and decision-makers to analyse future scenarios and to foresee bottlenecks and possible rehabilitation strategies.

5.2 Data requirements

The application of the developed techniques requires a dataset that involves collection of data for land use/cover from at least two different years in order to determine the changes in different land use classes. Other datasets required include ancillary maps that provide data to explain the process of urbanization or land use change, such as elevation, slope, boundaries, road network, or other services including gas, electricity and cable television, rivers, water bodies, etc. Depending on the availability of maps showing plans for future housing developments, business and industrial investments, and road expansion, these can be used to asses future scenarios.

Some of the constraints involved in using this modelling paradigm come from the limit on the amount of data required and the type of data. The implementation of satellite and remote sensing techniques has enhanced access to spatial information globally in recent years. Some issues still remain, for example the type of sensors used and the objective of the project are dependent on the time the information was capture and therefore influence the quality of the dataset. Also the classification of land cover/use for agriculture purposes differs from that used for urbanization. In both cases the built-up area can be identified, but the land use within the built-up area requires ground-truth verification points for instance commercial and industrial areas.

Other issues are related to the fact that by using remote sensors the available datasets refer entirely to the physical world. In other words, is possible to have information about the environment, boundaries and some infrastructure but it is not possible to map the decisions taken by political actors, decision makers, institutions, the economists, the business market, etc. All of which can be important in setting the rules that drive urbanization. The work presented here makes use of information and datasets that are currently freely available on the Internet together with other datasets that are acquired specifically for the case study.

For example, in the case of urban areas in Europe land use maps can be obtained from the European Environmental Agency corresponding to the Corine Dataset. The dataset exists in different resolution for the years 1990, 2000 and 2006. Since the focus of the study is on urban areas a higher level of detail is considered necessary for artificial areas and therefore the Corine land use classes were regrouped into 6 classes mainly bringing together the agriculture areas, forests and pastures into a class referred to as 'vacant land'. The other classes inside the urban agglomeration are: the continuous urban fabric which was renamed 'residential 1', the discontinuous urban fabric was renamed as 'residential 2'; and the industrial/commercial fabric was kept the same together with recreational land use. Airports, water bodies and construction sites were separate Corine land use classes and were regrouped into a single class called 'not modelled'. They were considered static classes in this study, that is, they did not change during the analysis.

Two different sources were used for the terrain data: Shuttle Radar Topography Mission (SRTM) data with 100 meter resolution and ASTER Global Digital Elevation Model (ASTER GDEM) data with 30 meter resolution. Both sources were used to match the resolution of the land uses acquired with the Corine Data. Both DTMs were projected and used to generate the slope raster maps for 100 and 30 meter resolution.

Other datasets representing the motorways, trunks, primary and secondary roads, rails, rivers and canals were downloaded from the open street map project (http://www.openstreetmap.org/).

5.3 Relation between water drainage and land use

The links between the spatial distribution of a drainage and other features or characteristics of an urban area has been previously studied by some researcher for example Mair et al. (2012) studied the similarities between roads, water distribution and sewer networks. The author's concluded that water distribution network can be generated from the road network. Blumensaat et al. (2011) presented a method to derive a drainage network layout based on the manipulation of the DEM and the street network; data available on the Internet is used to estimate configuration parameters to build a drainage network model. The link between change in land use and change of urban drainage networks has not been sufficiently studied or well documented. This may be due to the lack of appropriate historical records concerning the changes and/or construction done in the networks. Currently, there are many cities which have poor information on their existing drainage networks. It is often the case that the dimensions of pipes and ancillary structures and even their locations are not well documented. Given the continuous development of cities and their infrastructures, the proper management of the water infrastructure requires on-going investments in terms of the man-power, time, methods and available technologies which can be employed to capture and manage the data effectively.

A regular issue in water infrastructure planning and management is how to extend the network to cope with the growth of new urbanised areas so that the impact on the existing infrastructure is kept to a minimum within an overall budget. Further, since almost every intervention in the existing drainage network may disrupt other urban services and networks, the extension of the drainage network requires careful analysis. One of the main concerns regarding the planning process is the unknown characteristics or *state* of the system in the future. The use of methods and tools such as agent-based models and cellular automata for land use change can be used to understand the likely future changes in urban areas. These tools make use of spatial datasets that cover the urban area of interest.

The question about how to derive the connection between land use and the existing urban drainage network is addressed in the present work. The existing urban drainage network at any moment in time is the result of a series of decisions made in the past. These decisions reflect the impact of design rules, policies, the economic budgets, and unplanned actions as a whole. The current state of the drainage infrastructure also reflects the level of knowledge and how advance society is. In other words, is there a connection between land use and the drainage network apart of the estimation of runoff.

To address this question, we need to carry out a spatial analysis to see if there is a connection between the land use in an urban area and the properties of the urban drainage network.

The first step in this analysis was to focus on larger pipes, pipes with a diameter larger than 400 mm were selected from the drainage network layout. The second step consisted of the creation of 'buffers' or 'rings' around the pipes at every 100 meters for a total distance of one kilometre. This task was conveniently done in ArcGis 9.3, and enable us to explore the proximity of different land use classes to each pipe.

The area of each land use class in every ring was calculated leading to a knowledge of the distribution of land use classes along the main pipes. A similar analysis was carried out on the interception of the main pipes and with the land use classes using the zonal statistics analysis toolbox in ArcGis 9.3. This analysis produced a distribution of the length of pipeline that is intercepted for each land use class.

5.4 Methods to deduce the layout of the system

The development of an urban drainage network requires large investment by the community. Among the many factors that affect construction and operational costs are the diameters, installation depths, slopes, construction and operation of overflow structure and the use of pumping stations. As a basic principle, urban drainage networks are designed to follow the slopes of the natural terrain to make best use of gravity, and to minimize excavation costs and the use of lifting stations. The combination of these variables, the constraints imposed by the topography and the size of the system make it hard to analyse it manually and computational tools are therefore required. The layout of a drainage network is required in order to size pipes and ancillary structures. Researchers have addressed this problem differently in the past; for example Li and Matthew (1990) proposed a discrete differential dynamic programming DDDP approach to find an optimal solution. In their approach an initial layout is given using the graph theoretical Dijkstra algorithm, and the pipes are then sized. After that the variables of the network are fixed and the layout is optimized using graph theory. The steps are repeated in a cyclically until the objective function does not change in comparison with a threshold value. A similar approach also using graph theory is described by Diogo and Graveto (2006) and Haghighi (2012). In this study we start the optimization by deriving the initial layout of the network which is then optimized. The idea is to assess the impact of future developments on the existing network. Two approaches to obtaining the initial layout were tested. The first approach uses an agent based model that simulates distinct raindrops moving over the topography, under gravity. The second approach enhances the first approach by incorporating the road network as an influence on the direction of the network. The second approach was implemented in ArcGIS 9.3.

5.4.1 Approach 1 - Agent-Based Model

Agent-based models are often referred to as emerging paradigm models for which a population of autonomous agents interact to achieve a certain goal (see for example, Batty, 1997). In the present work, an algorithm that simulates rainwater drops as agents was implemented in Netlogo; this is an agent-based platform (Wilensky, 1999). The agent based model needs the digital elevation data and the movement rules for the agents. The agents interact locally and will move to a lower elevation from their current locations. Each raindrop or agent has an elevation as a property. Several agents can be stacked one on another at the same location until there is a sufficient number of them to generate a positive gradient from the location to the nearest minimum neighbouring cell;

in which case the agents move to the lower cell thereby tracing the trajectory or direction of flow. The agents are generated manually by the user using with the computer's mouse at particular points of interest derived from the previous step in the buffer analysis. The points use the most interesting land use clusters to define the position of the main pipes in the drainage network. of interest are derived from the analysis done in the previous step, basically defining the most interesting land use clusters to position the main pipes of the drainage system. As a result of the simulation an image emerges forming the pattern that describes a flow path.

5.4.2 Approach 2 Cost-weighted Raster

The 'Cost-Weighted Raster' model connects each of the generated attraction points in the drainage catchment formed in the study area with its outlet. Each link is treated as a straight line. The model creates a network layout taking into account the factors of land use, road alignments and terrain slope. The Euclidean distance between two points may be modified due to constraints that are the result of these factors. The constraints result in restrictions between certain cells and encourage connections to some other cell(s) of interest. Each link in the network must have a positive gradient. We also showed by a separate statistical analysis that the network should be aligned where possible with the primary and secondary roads. We also concluded that when extending the layout to new areas it may be cheaper to cross vacant land in order to have the shortest distance to the existing system.

This model was developed in Model Builder and Python Scripter. The model used the land use map simulated by the CA model, the DTM of 30 meter resolution and the shape files of motorways and trunks, primary and secondary roads. The DTM was used to calculate the slope map of the study area.

The Hausdorff method for the distance between the generated layout and the existing system was used to measure the fitness of the generated layout. The Hausdorff distance is the longest distance between one set of polylines and another (Hangouet 1995)

5.5 Case Study Birmingham, UK

The information regarding the urban drainage network was obtained through the SWITCH Learning Alliance established in Birmingham (www.switchurban.eu). The hydraulic model of urban drainage network used in this study was obtained from Severn Trent and it covers the Upper Rae Main catchment. Previous research done with this dataset has been reported by Thuy (2009) and Last (2011). The other information and datasets for this case study was presented in detail in chapter 3.

5.5.1 Urban Drainage Model

The model of urban drainage network for the city of Birmingham consists his corresponds to two catchments the Upper Rea Main and Griffins Brook areas, in the south west part of the city. The model was built in Infoworks and from there was transformed into EPA SWMM. Since the conceptual modelling framework of both tools is different it was necessary to calibrate the model in SWMM, assuming that the outcome of infoworks correspond to the reality or observed data. Only a small part of the model was available in the study area. The model consisted of 2796 sub catchments and 6340 conduits; figure 5.1 shows a layout of the drainage system.

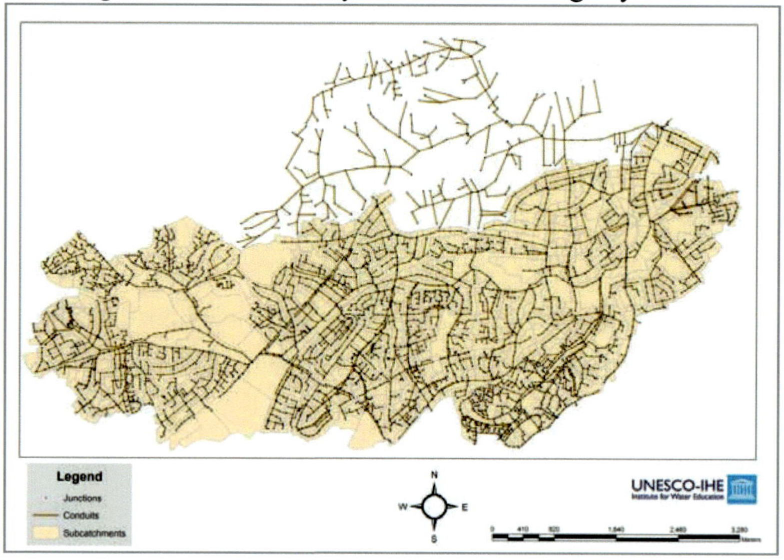

Figure 5.1. Full Model Urban Drainage System - Birmingham.

Table 5.1. Characteristics Full Model Urban Drainage System - Birmingham.

Item	Value
Rain gauges	8
Subcatchments	2796
Junction Nodes	6365
Outfall Nodes	40
Conduits Links	6340
Orifice Links	49
Weir Links	18
Outlet Link	1

A total of 15 models were built in EPA-SWMM in order to calibrate the hydraulic model of Birmingham. In these models was changed basically the values of Manning's overland flow coefficients and the parameters of infiltration. The best results were obtained with the model named EPA-SWMM_15, and this was selected to improve the calibration results; from model 15, four sub-models were created, on these the time step was changed in order to obtain a more stable model.

To calibrate the models, as observed values in SWMM were assumed the values reported by Infoworks. The results of the calibration process are shown in figure 5.2. From the analysis of these results it can be observed that the calibrated network is able to reproduce the pattern of the flow as well the total discharge, with correlations above 99%.

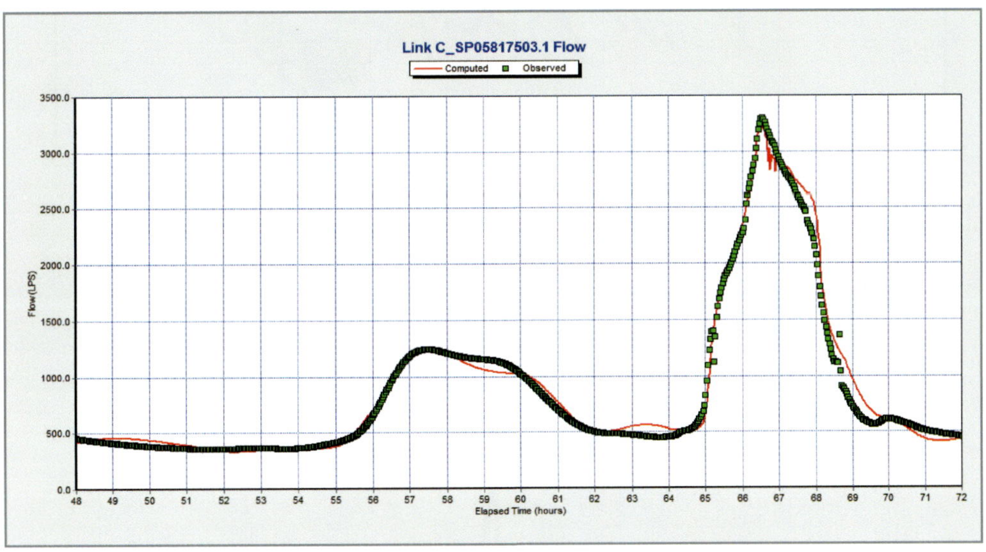

Figure 5.2 Calibration Result Full Model

5.5.2 Pruned Network

Since the emphasis of the study is expanding the drainage network to future developments the model was pruned to decrease its complexity. The pruned network contains pipes with diameter 400 mm or larger. The scheme of this pruned network is presented in figure 4. This pruned model contains 1229 conduits. The simplified model was also calibrated to adjust the parameters in order to reflect the transport capacity of the full model.

The scheme of this pruned network is presented in figure 5.3 and the calibration output is given in figure 5.4. The pruned network was more convenient for land use simulations as a driving factor. Also, it was used as a basis for the analysis of sewer network expansion described in the following section and for the optimal design of sewer network that accounts for urban expansion

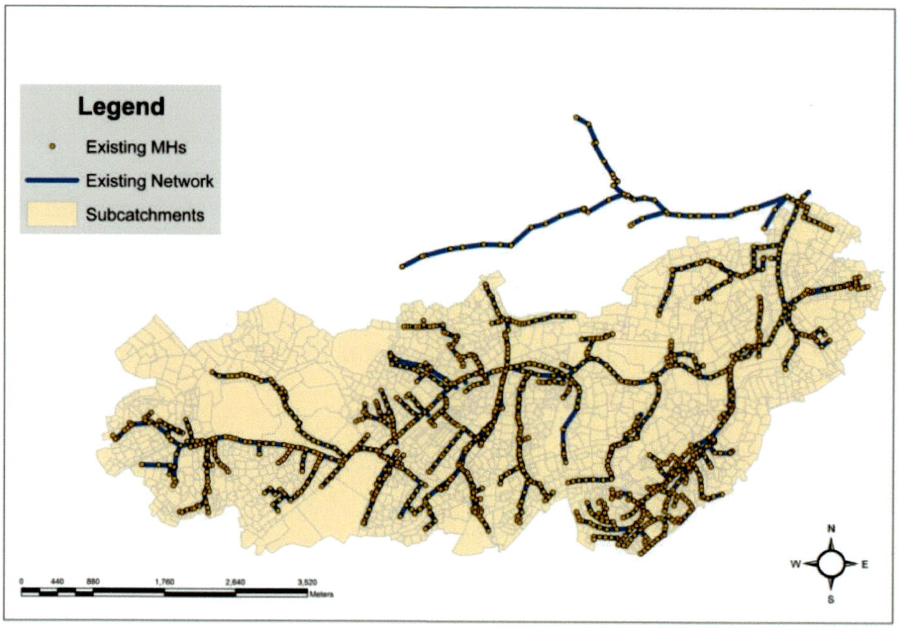

Figure 5.3. Pruned Drainage network for the study area.

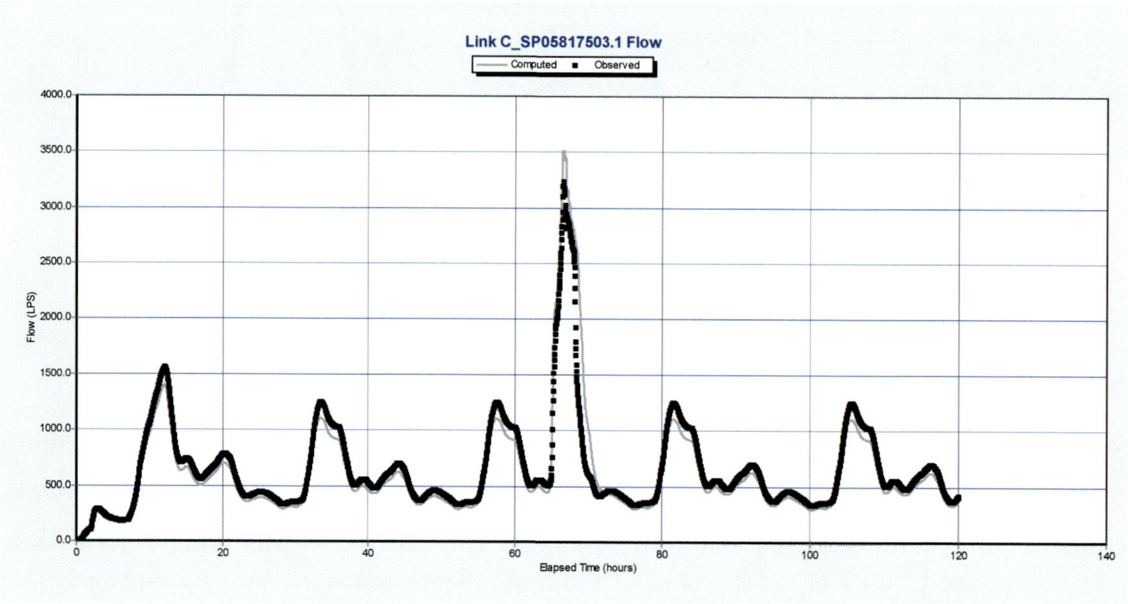

Figure 5.4. Calibration of the pruned model.

5.5.3 Land use and urban drainage system

As mentioned above a corridor analysis was done along the main drainage pipes to derive a connection between the land use and the existing urban drainage network. The GIS layer with the buffers or corridors was then overlayed on the land use map for the year 2000, see figure 5.5. The outcome of this analysis is a distribution of the land use area per category that falls into each corridor. Figure 5.6 shows the result of the analysis for the land uses: 'residential 1', 'industrial/commercial', 'parks' and 'recreational'.

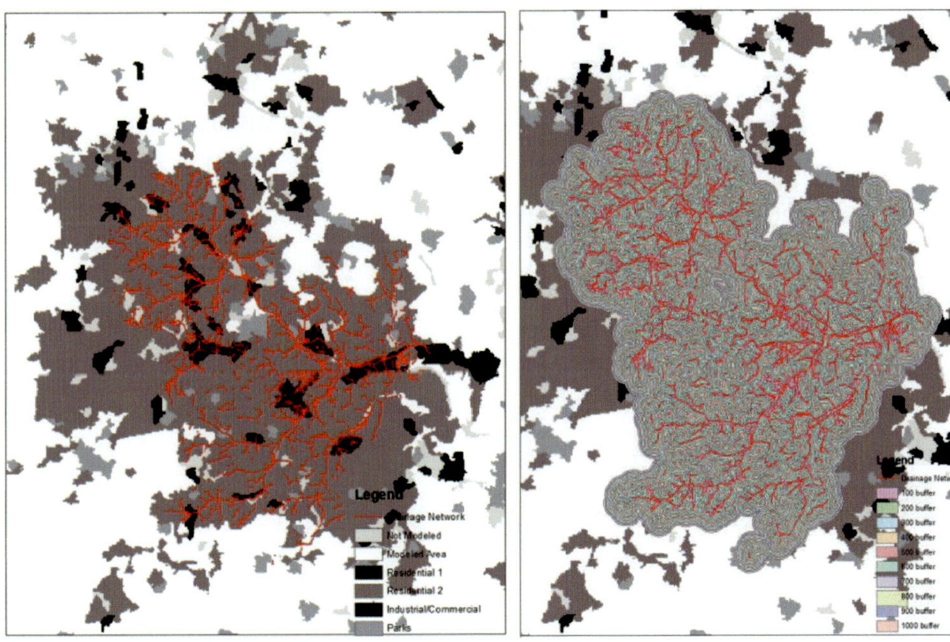

Figure 5.5 Corridor analysis along the main sewer drains

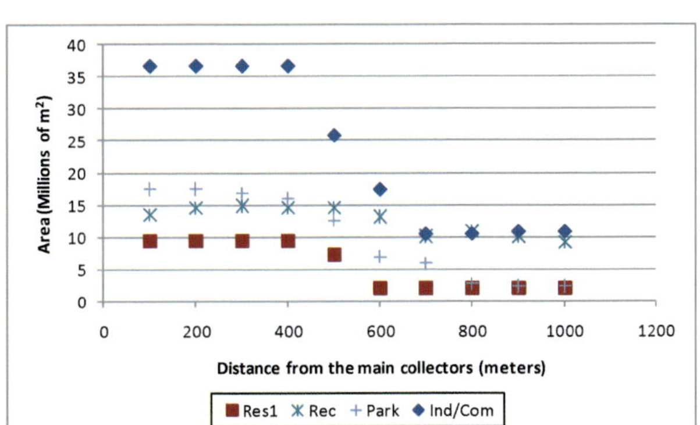

Figure 5.6 Land use area distribution with distance from the main pipes

Information obtained from the corridor analysis can be used to anticipate the location of new pipes in futures scenarios based on the estimated land use. The land uses industrial/commercial- and residential 1 show a significant change in the intercepted area at about 500 meters. This means that the main collectors of the network are positioned within 500 meters of these land use categories. It is possible that this finding reflects the traditional engineering practice of using a higher return period in the design and construction of drainage networks inside urban areas to provide a higher degree of safety to valuable parts of the city including governmental institutions, hospitals, industry, etc.

For the land use class 'residential 2' the analysis showed a homogeneous distribution of the area within the buffers, and therefore it does not provide valuable information about the location of the main drains.

A similar spatial analysis was carried out by overlaying the raster map of the land use for the year 2007 and the shape file containing pipes with diameter greater than or equal to 400 mm. Using a zonal statistical analysis method it was possible to identify the total length of pipes for each land use (see Figure 5.7).

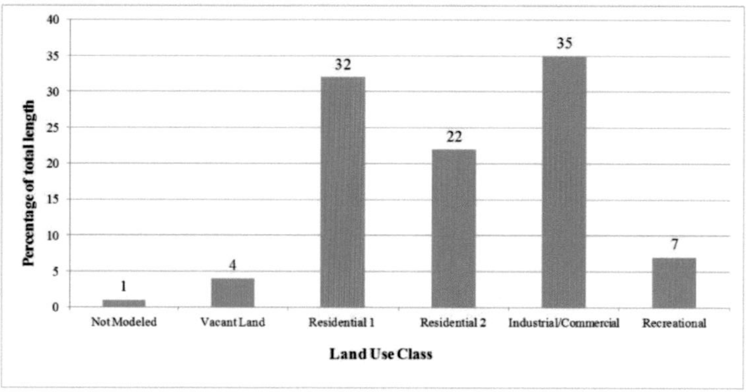

Figure 5.7. Percentage of total length per land use class

The results presented in Figures 5.6 and 5.7 give an indication of the spatial position of the main drainage pipes. In Figure 5.7, the land use class 'Residential 2' shows that a high percentage of pipes are intercepted. Also, by observing Figure 5.6 we can note that the 'Residential 2' land use is homogeneously distributed in space. However, it does not provide information about the positioning of the main pipes. With this information rules can be defined to set up the behaviour of the agent-based algorithm to extend the drainage network.

Another important aspect of urban drainage is that the drainage is largely determined by the topography. In fact, this is the main criterion used for the design of urban drainage networks. Figure 5.8 presents a comparison between the existing urban drainage network with pipe diameters larger than 400 mm in Birmingham and the derived natural drainage network obtained by processing the DTM. The flow accumulation map was used to derive the natural drainage.

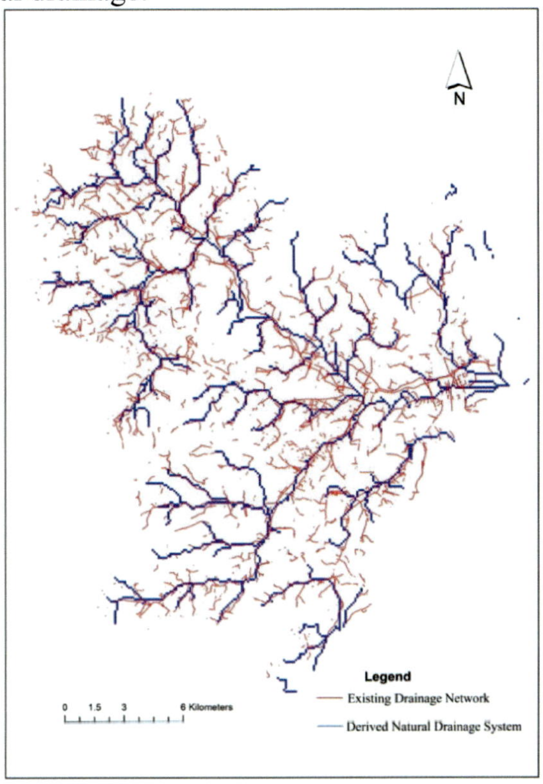

Figure 5.8 Existing urban drainage network and derived natural drainage network

Figure 5.8 shows a relatively high spatial correlation between the constructed drainage network and the GIS-derived natural drainage network produced for the terrain topography. The estimated spatial correlation using the fuzzy similarity comparison method of Hagen (2003) is 0.74.

The spatial analysis shows that it is possible to relate information from the land use map and the topography (elevation and/or slope) with the existing drainage infrastructure.

5.5.4 Deriving the network layout for existing system

To test the approach an algorithm that simulates the dynamics of drops of rain was implemented in Netlogo. This agent-based platform uses the same digital elevation model and the points of interest according to the classification of land use and the corridor analysis discussed above. The points of interest were derived as the centroids of clusters with the land use ('residential 1' and 'industrial/commercial'). The result of this simulation is a pattern that describes a flow path. Figure 5.9 shows the results of the simulation compared with the existing drainage network.

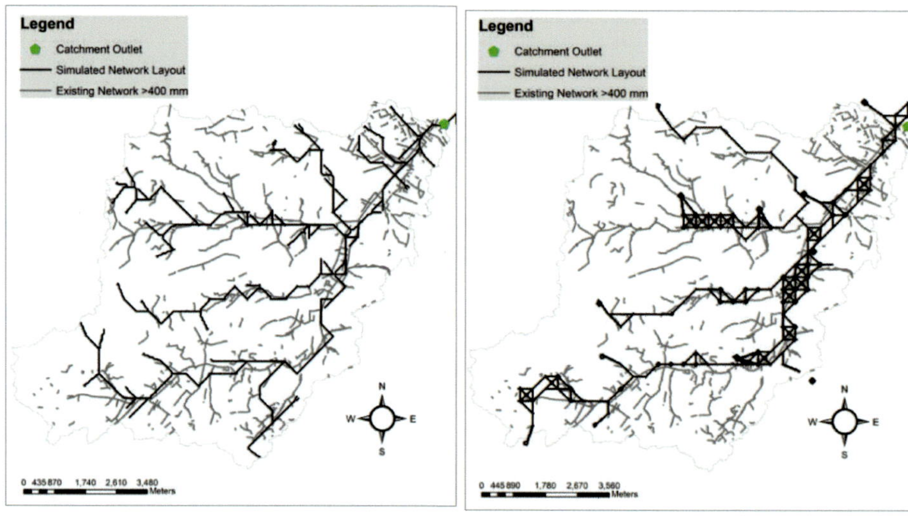

Figure 5.9 Layout of the drainage network derived with an agent-based model for 30 (left) and 100 (right) meter DTM cell resolution.

As can be observed from Figure 5.9 the algorithm simulating the dynamics of the rain drops can derive the layout of the existing main drainage network reasonably well. This shows that by combining the information on the land use (i.e., the points of interest) and the topography of the area a good approximation can be obtained for the layout of the main pipes of the network. Figure 5.9 also shows that some areas were not covered by the network. This is due to the fact that there are no land use clusters of interest in these areas and there is a need to find other key information in the dataset that can contribute to filling this gap.

Information regarding land use, topography, slope and the road network, was combined to create a cost-weighted raster in order to improve the layout generator. The cost-weighted raster is a map that contains a cost factor for each cell; the factor is used in combination with the start (new manholes) and end points (closest existing manhole) the links between the attraction points. With this map it is possible to identify the least cost path (the one with the minimum total cost) from the starting point to the nearest manhole.

The creation of the drainage layout is achieved by crossing areas with gentle slopes, following the alignment of the existing roads and passing through 'vacant land', 'residential 1' and 'industrial/commercial' land uses. The outcome of the simulation for both cell resolutions (30 and 100 meters) applied to the Upper Rae Main catchment is shown in Figure 5.10. This comparison of the results with the existing drainage system proves the validity of the method.

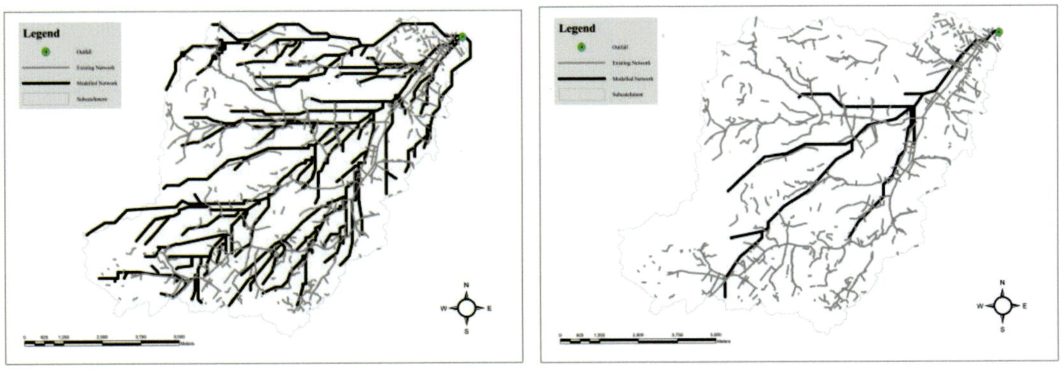

Figure 5.10 Drainage network layout generated for 30 (left) and 100 (right) meter cell resolution.

Figure 5.10 shows that the layout has a good approximates well to the layout formed by the main drainage pipes in the Upper Rae Main catchment. The Hausdorff method is used as a measure of fitness as indicated in the methodology. The results obtained are summarized in Table 5.2.

Table 5.2: Hausdorff distance for the generated sewer layout.

Cell Resolution	Netlogo Model	Cost Weighted Raster
30 m.	124.92	169.85
100 m.	115.7	152.59

Figures 5.9 and Figure 5.10 together with Table 5.2 show that the resolution of the raster used to predict the layout is important in the final result. A better spatial coverage of the drainage network is achieved when the resolution is increased. Some areas of the catchment are not covered by the simulated layout; this is likely to be due to the cell resolution, to a lack of information and/or to a limitation in the method. The outcome of the Hausdorff distance calculation shows that the maximum separation of the derived and existing networks is within 100 to 200 meters range. While this indicator provides a measure of average proximity between the two layouts, there are significant differences in coverage and also in the direction of some of the line segments. This also suggests that a different indicator needs to be formulated to assess the fitness of the derived layout. In an attempt to solve this issue the area of separation between the two networks is propose as an indicator. The calculation was done using Arcgis, delineating manually the polygons between the two sets of polylines, and then calculating the total area. Figure 5.11 shows an example of this area-separation method for the simulated network using the cost weighted raster with a resolution of 30 meters. The maps for the

simulated layouts presented in figure 5.9 and 5.10 with estimated area of separation are presented in annex4.

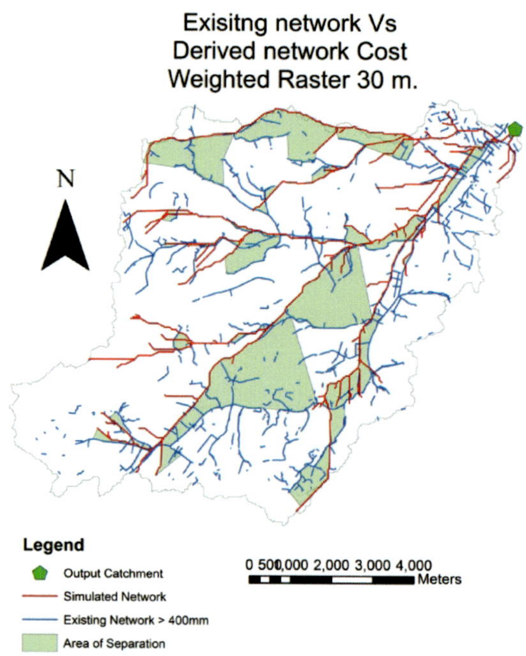

Figure 5.11. Area of Separation simulated layout with cost weighted raster 30 m.

The estimated areas of separation for all the generated layouts are presented in figure 5.12. The calculation of the area of separation can be automated in the future to avoid the manual intervention and judgment in the calculation of the indicator. A couple of procedures were tried within Arcgis 9.3 but they fail mainly because the existing system is a selection of pipes of the whole network; therefore there was not continuity in the set.

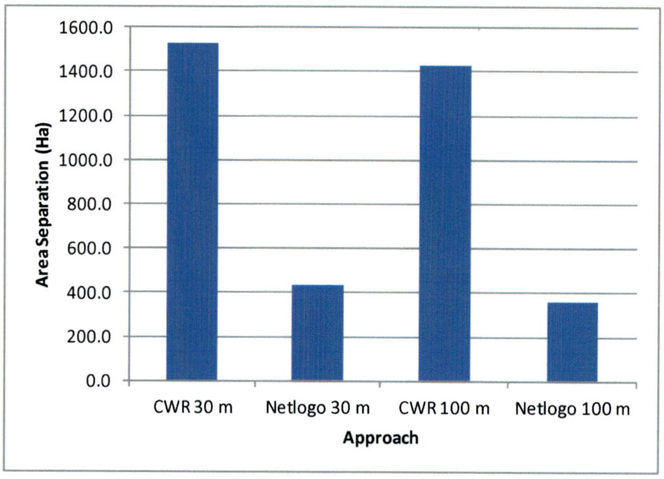

Figure 5.12. Area of separation for the generated layouts

Figure 5.12 shows that the indicator based on the area of separation is more sensible than the Hausdorff distance presented in table1. The values of the Hausdorff distance are more homogenous, hence is difficult to clearly say which approach is better. The area of separation can better account for those lines that are not accurately position in space, like those predicted by the cost weighted raster.

The range of separation calculated with the Haussdorf distance can be explained by the fact that the method is guided by the topography and tries to mimic the natural drainage paths, while in reality the exiting manmade drainage system follows the topography and it is not necessarily constructed along the natural drainage network but adjacent to it. The use of the point of interest found with the corridor analysis can be even more valuable in zones with flat topography where the natural drainage flow paths are difficult to delineate.

When all the centroids of interest have been created and connected to the outfall of the catchment, they are then converted into segment lines and as such they represent the conduits or pipes in the urban drainage network. Similarly, the points created at both ends of the lines represent the manholes. One of the advantages of this approach is that all information is stored in a geo-database and this allows input files for the SWMM model to be created.

This approach can be further extended and used to derive the future layout of the network by using as an input the land use change model.

5.6 Extending the drainage network layout to new developments

The approach to derive the drainage network layout that was tested with the existing systems was used to generate the future layout of the network. This was done considering the land use changes generated by the urban growth model for the year 2040 that were described in chapter 3. Figure 5.13 shows the layout of the network considering the future land use map of year 2040 for model M1 with 30 meter resolution and figure 5.14 presents the result for the model M2 with a resolution of 100 meters.

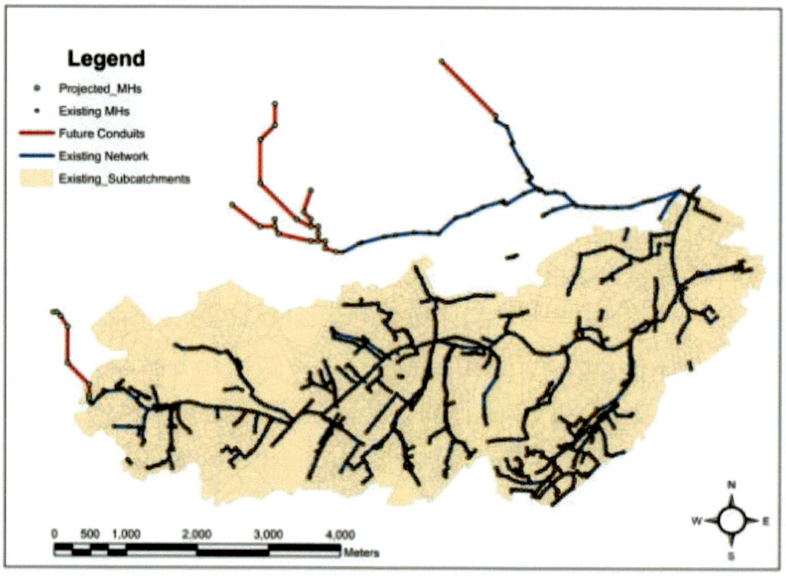

Figure 5.13. Expansion of the drainage system for scenario 2040, model M1 (30 m. cell)

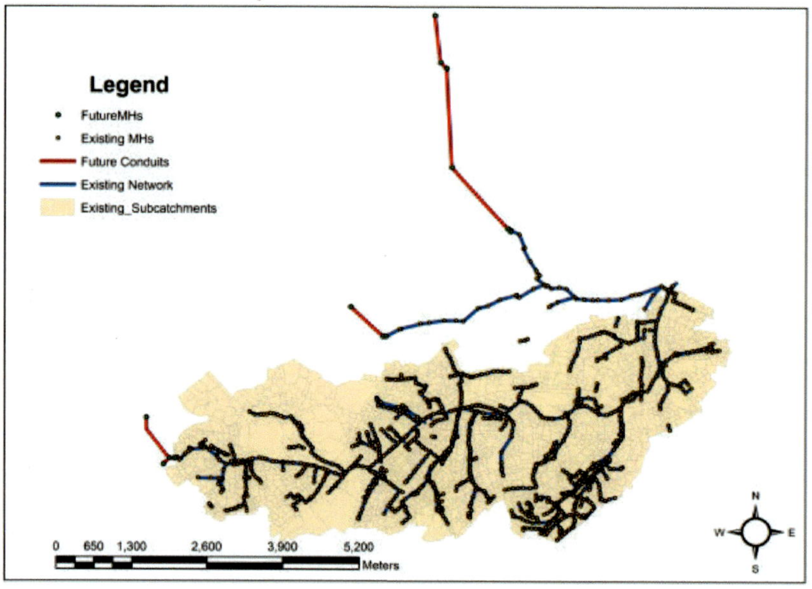

Figure 5.14. Expansion of the drainage system for scenario 2040, model M2 (100 m. cell)

The generated layouts considering the outcome of the two land use models show that the cell resolution is important to obtain more detail layout if the drainage system. It is also worth to mention that even though the land use model use a different cell resolution the spatial location of the likely future expansion of the drainage network are the same. To assess the impact of the future land use change on the existing infrastructure the layout produce with the land use model M1 is used for the purpose of design and interconnection.

5.7 Impacts of urbanization in the existing infrastructure

5.7.1 Expansion of the drainage network

For the future expansion of the drainage system the design of the conduits was carried out by using the framework developed by Sanchez, 2007 and presented in more detail in Sanchez et al 2008. The framework consists of an optimization algorithm (NSGA II) coupled with SWWM that assess the capital cost of the system as a function of the length and diameter of every element and the estimated damage as a function of the volume of flooding (see figure 5.15). Out of the Pareto front that is generated by this multi-objective optimization the cheapest solution that produces no flooding is selected as the best option for design purposes.

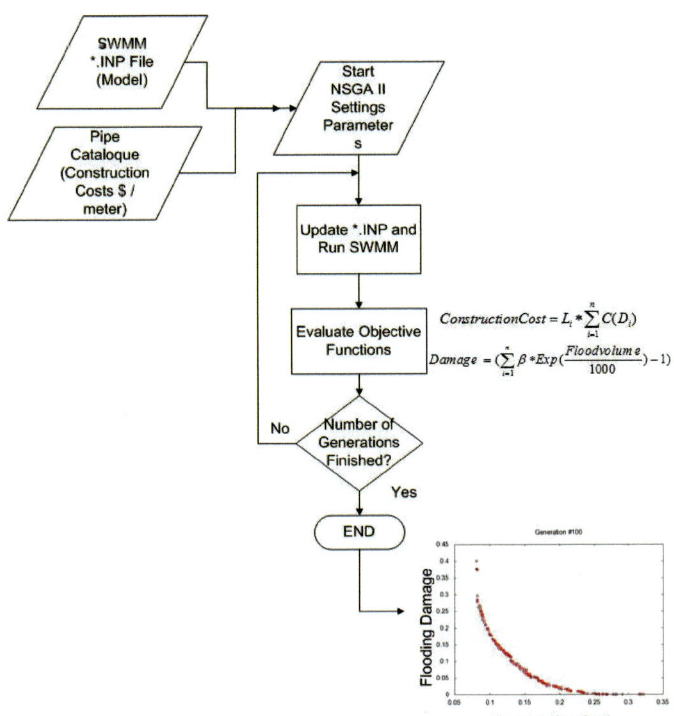

Figure 5.15. Optimization loop to design the expansion of the drainage system

Once the elements of the future subsystem expansion are design, they are connected to the existing system or model to assess the impact of the expansion into the existing infrastructure. The assessment is performed by considering the volume of flood and the duration of the flooding in the new scenario that considers future land use changes.

In order to assess the impact of urbanisation on the existing urban drainage infrastructure, the future layout of the system was analysed. Knowing the spatial position of the future drainage layout, the ArcHydro tools were used to delineate the sub-catchment areas or areas that will naturally drain towards the drainage conduits. The sub-catchments were overlayed on the future land use map for the year 2040 to obtain the distribution of land uses. The runoff extractor tool was then used to estimate the percentage of pervious and impervious surfaces. Figures 5.16 and 5.17 show the surcharged nodes map for the pruned existing system and the expanded model considering the new developments. Both models were run for a recorded event with a return period of two years and 60 minutes duration.

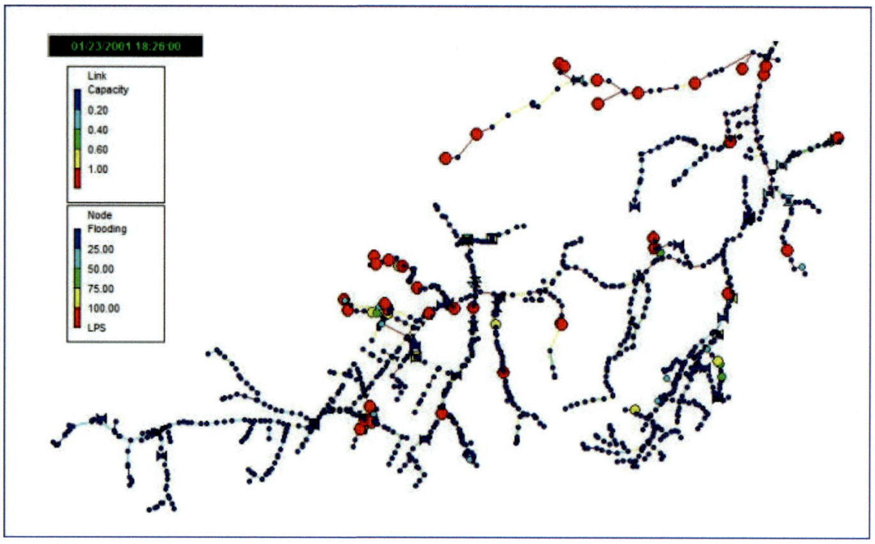

Figure 5.16 Surcharged nodes map for the pruned existing system.

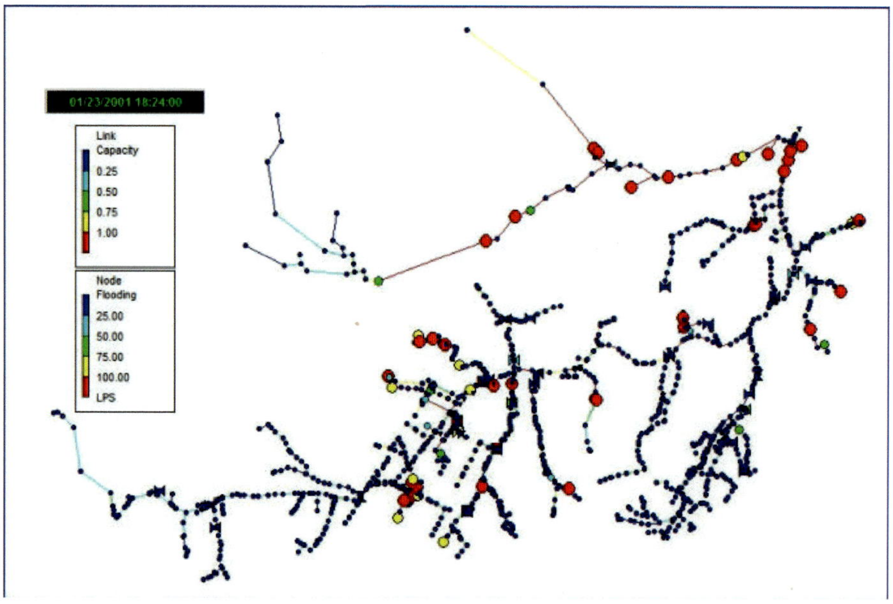

Figure 5.17 Surcharged nodes map for the expanded system considering new developments.

The total area with new developments is 683 Ha and is distributed over 32 sub-catchments. The average percentage impervious surface is 0.17 and the estimated average values of the manning n-impervious and n-pervious are 0.002 and 0.38, respectively. The estimated total runoff contribution of the new developments is 20940 m^3. Table 5.3 presents some indicators to assess the performance of each system.

Table 5.3 Drainage system performance indicators for new developments.

Indicator	System	
	Pruned	**Pruned Extended**
Total Flood Volume [m3]	19301	36207
Hours Flooded	338.17	395.93
# Nodes Flooded	221	271
Links Surcharged	360	366

Figure 5.16 and Figure 5.17 together with the values presented in Table 5.3 show that the expansion of the drainage network will increase the total number of locations with flooding to 50 nodes or manholes. Part of the new developments is likely to occur in the area where the system is already in a critical condition (i.e., does not have sufficient capacity) which is highlighted by the total volume of flooding increasing to almost double its original value while the total duration of the flooding does not change significantly. Out of the total volume of runoff that is generated in the new developments (20,940 m^3), a great portion (16,906 m^3) will not be conveyed by the pipe system and will exacerbate the potential flood-related problems. This can also be observed in Figure 5.18 which shows the peak flow at the outfall of the system. The outfall is at the entrance to the wastewater treatment facility.

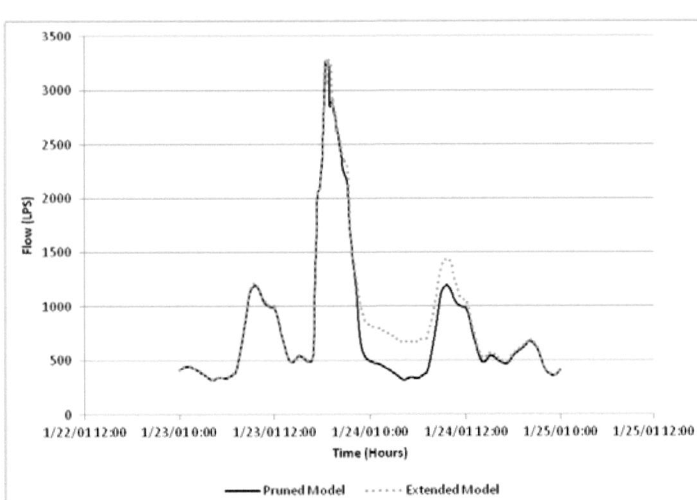

Figure 5.18 Comparison of discharge hydrographs at the outfall of the system (pruned-existing and expanded model network).

Figure 5.18 shows a significant change in the flow at the outfall pipe for the tail of the event. This shows that it will take longer to reach the average operational condition of the treatment facility. This has implications for the operation of the system as the volume of tanks in the Wastewater Treatment Plant needs to be carefully considered and analysed against CSO (combine sewer overflow) frequencies and volumes in the future.

The total number of surcharged pipes does not change significantly considering the scenario for the year 2040. Table 5.3 shows that 360 pipes are surcharged in the system. Surcharge is the ratio of the maximum peak flow for this particular event and the full flow capacity of each pipe. This also identifies those pipes in the network that have pressurised flow conditions. Out of all the surcharged pipes, those elements with the largest change (i.e., pipes for which the increase in surcharge capacity is greater than 0.5) were identified. These pipes are presented in Table 5.4.

Table 5.4 Pipes with the highest change in hydraulic performance

Priority Pipes	Pruned System			Extended Pruned System			Increase in surcharge
	Flow (LPS)	Velocity (m/s)	Max/Full Flow	Flow (LPS)	Velocity (m/s)	Max/Full Flow	
C_1_360.1	954.54	2.16	1.59	1392.02	3.15	2.31	0.72
C_1_370.1	984.11	2.23	1.67	1301.91	2.95	2.21	0.54
C_37_250.1	1165.85	1.83	1.73	3296.98	5.18	4.9	3.17
C_SP01773508_W.2	2189.67	4.96	1.73	3006.31	6.8	2.37	0.64
C_SP05803501.1	667.4	1.44	0.89	1165.91	1.56	1.55	0.66
C_SP05806454.1	55.57	1.44	1.58	92.12	2.32	2.61	1.03

The identification of these critical conduits is useful information that can be used to evaluate potential rehabilitation strategies for the system. This again can be undertaken as a dynamic optimisation problem (see for example, Sanchez, 2007 and Vojinovic et al., 2006).

5.8 Conclusion

The method presented in this chapter demonstrates that connecting the distribution of land use in an urban area with other urban infrastructure information such as roads, canals, and drainage networks can yield valuable information that can be used to derive the key rules relating the different factors and to obtain a good approximation of the drainage network layout. Two techniques were used to derive the layout of the system, one using agent-based models and the other one using similar tools to build as a set of raster operations within ArcGIS. The approach was tested on a catchment in Birmingham, UK. The results show that both techniques performed well for a scoping analysis at an urban scale level to derive the main pipes of the drainage network. As anticipated, the quality of the information affects both techniques. In particular, it was found that the cell size of the elevation map plays a major role.

In the case study area it was found that the main collectors of wastewater pipes are located within 500 meters of the land use classes 'residential 1' (Continuous urban fabric) and 'industrial/commercial'. These are the main land use classes that influence the location of pipes in future scenarios.

The case study results show that the application of a cellular automata technique for simulating urban growth processes can yield promising results. The spatial analysis identified the need to model the internal changes in the city land use and not only the expansion or contraction of the urban core. The development of the case study encountered several limitations regarding the nature and quality of the information needed to setup the model. The calibration process of the cellular automata model was undertaken despite the fact that not all the variables that are involved in the urbanization phenomena were used. The work performed highlights the need for a more multidisciplinary analysis.

To visualize the impacts of future urbanization growth on the existing drainage infrastructure, the output map of the cellular automata model for the year 2040 was used to derive a possible layout of the system in the future. The possible expansions of the drainage network were designed on the basis of the future land use map. The catchment parameters needed for the rainfall-runoff were estimated with the runoff extractor. The pipes were sized using the NSGA II algorithm coupled with SWWM within an optimization loop. The new pieces of the drainage network were then interconnected to the existing model of the drainage system to assess its new performance and to evaluate the consequences of the future land use changes for the existing infrastructure.

The interconnected model for the future urban growth scenario of Birmingham shows that the future developments will contribute further to the flooding problem if no improvements are made to the existing drainage system. The total number of flooded manholes will increase by 50 and most of the runoff volume that will be generated by the new developments will exacerbate the flood-related problems due to the lack of hydraulic capacity of the existing system. The approach presented in this paper can also be used to identify the critical pipes that require immediate attention for rehabilitation purposes. The same approach can be used to evaluate rehabilitation strategies to improve the performance of the system now and in the future.

6 Framework to model cities future water infrastructure

6.1 Introduction

Cities around the world are currently facing considerable pressure to cope with urban development and economic growth. On the one hand urban infrastructure is aging and there is a rapid growth of population while economic resources are in short supply. Many cities changed from being small and isolated to becoming large, economically interconnected urban centers. Urban sprawl generates a loss of natural vegetation and open space, affecting the continuity and connectivity of natural habitats and eco-systems. It is estimated that 100 years ago 15% of the world's population lived in cities, while the present situation is about 50%. During this period of time the number of people living in cites has increased 100 times; USGS 1999. This is generating considerable stress in the surrounding environment and the population itself. Even though the urbanization process has been occurring for centuries, still the temporal and spatial drivers and variables that shape this phenomenon are little known or well understood. Under such circumstances urban planning is very important to improve the effectiveness of the investments and interventions that take place in the urban system.

In the United States, the EPA estimates the cost of the capital investment to maintain and upgrade drinking water and wastewater treatment systems in 2010 was $91 billion, out of which, only 36 billion were funded. Thus, there was a gap of $55 billion for that year alone; ASCE 2011. The results of EPA's 15th national survey of capital costs to address water quality or water quality related public health problems showed that the total wastewater and stormwater management needs for the nation are $298.1 billion as of January 1, 2008. This amount includes $192.2 billion for wastewater treatment plants, pipe repairs, and buying and installing new pipes; $63.6 billion for combined sewer overflow correction; and $42.3 billion for stormwater management. Small communities have documented needs of $22.7 billion. Considering that the water infrastructure is aging and that investment is not able to keep up with the need, there is clearly a need to increase the effectiveness of the investments, planning and rethink the urban water cycle not only in the US but elsewhere.

Urban drainage systems are often neglected by decision makers for continuing investments to upgrade the systems and keep them functioning well. Often, this is because the physical components of the system are not visible, and the hygiene and sanitation conditions in cities have improved, in particular in the developed world. Therefore, the problems related to urban drainage systems gain visibility when there has been a disaster. Considering urbanization growth and the uncertainties of climate change, problems are escalating in frequency, magnitude, and are more complex. Urban drainage networks can be considered as socio-technical systems where the actions taken in one part of systems can propagate to other parts with unforeseen consequences.

There is little understanding on how the social part is functioning as a system. Recently, with advances in remote sensing, there is an increasing quantity of data available to help understand how society is working, how to improve decision taking, how to use infrastructure better. The technical side of the urban drainage is relatively well understood, but the combination of the social and technical part of society is still not well investigated. Nevertheless, when considering climate change and its uncertainties in the future there is a big need to do things differently in order to strategically adapt to these changes. The adaptation strategies need to be well analyzed and conceived in a holistic way for two reasons. Firstly, the existing systems are the result of the collective behaviour of society, and it imposes constraints on future developments. Secondly, any intervention to rehabilitate or upgrade an urban drainage system is expensive and can disrupt other services.

Integrated urban water management is a challenging issue that aims at the sustainable use of the available water resources. The current practices in the sector are leading towards a crisis that is calling for innovative thinking and the adoption of new strategies, including integrated thinking and planning. There is also a need to develop tools that allow such integration.

In the literature there are a number of attempts to model the urban landscape and the changes in land-use for future development according to scenarios. The use of the cellular automata technique to model land-use changes is an example that has been successfully applied and calibrated to real life cases in Europe and different cities around the world, as presented by Barredo et al, 2003, Engelen et al, 2007, Soares et al, 2011, among others.

Urban planning is therefore very important to improve the effectiveness of the investments and interventions that take place in the urban system. The proposed approach aims at the integration of cellular automata models with physically based hydraulic models of the networks to determine the water infrastructure and performance in delivering adequate water services in the future and how this can shape the urban development process.

The integration of these models allows the exploration of several planning scenarios to assess the impact of certain actions, policies, regulations and to explore several urban futures. The result is a new approach for urban water infrastructure planning which helps water companies and municipalities to have a dynamic planning tool to improve the effectiveness of their investments and to be more environmentally efficient.

6.2 Data Requirements and processing

Like any model, the accuracy of the simulations is directly related to the data being used. It is important that available information comes from the same source and that the methodology to generate the data remains consistent over time in order to make comparisons objective. In this research, data that was freely available on the Internet was used. These data sets are becoming more reliable and are more accurate with time; still, care must be taken in analysing the data, especially data from different sources, taking into account the amount of data and the integration of the whole database. In this research it was evident that there is a lack of historical databases for urban infrastructure, land-uses, water services and road expansions. This is important because understanding our past is important to identify the causes and decisions that lead to the infrastructure we are using now, and this in turn has huge implications for the future.

6.2.1 Land-use maps

The existence of land-use maps is the key factor to be able to apply the proposed approach in this research. There is a need to have spatial databases with interpretations of main land-uses where the main features can be recognized such as urban extent, transportation networks, water features, main eco-systems and other important features. As the corridor analysis has shown in the two case study areas analyzed in chapter 4 and 5, the identification of internal land-uses such as commercial, industrial areas, parks and recreational facilities as well as the location of public buildings are important to identify patterns that help in the positioning of main water pipes. This type of land-uses required the use of different data sources than remote sensing. Ground true points or verification are required to define these land-use classes. The use of census information or the combination with tax office databases and municipality cadastre databases is useful in acquiring and continuously updating this type of land-use classes.

The use of mobile phone data use within the urban area can help in the identification of land-use clusters. By analysing the daily pattern of cell phone activity some industrial, commercial, office buildings, discotheques, restaurants, etc can be identify. Moreover, the use of cell phone activity can show the global movement of people in the urban environment and this is important to analyse and understand how the infrastructure is used in daily life.

Another important data source for the internal urban land-uses is the use of information from navigation systems that is probably update more regularly. Nevertheless, the challenge still persists in how to represent point information like the one from the navigation system with a grid kind of representation.

Land-use maps from the past are important to assess the pattern and trends of land-use changes. If these maps were not previously produced, the mains sources of information will be satellite images aerial photographs.

6.2.2 Understanding land-use change

The analysis of a temporal database of land-use maps is helpful to identify spatial patterns, rates of change, trends, key features and moreover insight on how a certain city has evolve under varying socio-economic and environmental conditions. Equally important, is to have access to such temporal database of maps is the understanding of regional development and history, especially in more interconnected cities or urban areas. Along with the land-use maps for several years, information regarding population data, historical analysis and timelines connecting historical events like in a forensic study will be helpful to explain the changes that are observed. This requires a multi-disciplinary observation of the events to construct a holistic analysis.

It is also important to have good information about the environment in the urban area and in the region; a good analysis of topographic features, climate, hydrology, adequate supply of water, sanitation and hygiene can encourage or limit growth and development. The development of the road network and transportation facilities plays a key role in urban changes. In the 18th and 19th century main urban areas were located along waterways and ports. With the development of the car and the road network many developments occurred along the road network. These patterns still drive urban development since connectivity plays a key role in the economy nowadays.

There is also a need to understand land-use change at regional scales. Urban developments have different patterns in different regions. In Africa, the urban primacy still dominates the growth; this is one main city, often the capital of the country, dominating urban growth. In Asia there is regional pattern with the growth of metropolitan areas with small cities being interconnected by good transportation systems. European cities tend to be more compact and there is very little growth of the urban extent; many changes are happening within the already built-up area or even shrinking the urban extent.

For example, according to UN Habitat 2008, settling near large bodies of water has clearly been an important factor in the economic and demographic growth of cities. Cities located near the sea have an obvious advantage: they provide access to sea trade routes and links. Globally, coastal zones are the most urbanized ecosystems, with 65 per cent of their inhabitants residing in urban areas. Cities located inland are now growing faster around the world than cities in coastal zones. Globally, cities located in mountainous regions grew at almost the same rate as cities located in coastal zones (approximately 2.5 per cent a year) between 1995 and 2000.

6.2.3 Modelling land-use change

The historical land-use patterns established with at least two land-use maps from different years, together with a set of maps such as transportation networks, services, topography and environmental features are used to explain the urban changes. Knowing that there is a lot of uncertainty in the datasets and models to simulate future land-use, the presence of certain features are used as an enabling variable for urbanization. In this approach a cellular automata model is used to simulate land-use changes based in statistical methods to calculate probabilities of urbanization.

6.2.4 Assessing the impact of land-use change

Modelling land-use changes provides a tool to explore and assess several scenarios for urban growth or development. Planners can use urban dynamics data to evaluate environmental impacts, test certain policies, delineate urban growth boundaries, zoning policies, and sense the magnitude of future water infrastructure requirements. By evaluating several trends associated with land-use changes over time, several hypothesis and/or solutions to solve bottleneck problems in water infrastructure can be analyzed ahead so that a better plan and hopefully a cost-effective alternative can be implemented.

6.3 Modelling Approach

In the modelling framework of this research the idea is to use the outcomes of Dinamica EGO and to incorporate relevant information from the urban water related infrastructure. In particular, the objective is to design the water distribution and drainage networks in the urbanising areas of a city based on the characteristics of the existing networks, and to rehabilitate the existing networks such that the whole urban water system is sustainable for the future.

The main components of the modelling approach are shown in figure 6.1:

A regional model: to construct scenarios for the distribution and allocation of land-use demand.

Cellular model: Computes the potential of land-use change and future land-use map

A numerical model: that describes the performance of the water systems (water distribution and drainage system).

A reverse engineering model: that takes the predicted new developments land-uses to assess the layout of the water services networks in such a way that the water services accessibility maps can be updated and create dynamic conditions that are incorporated in the simulation loop.

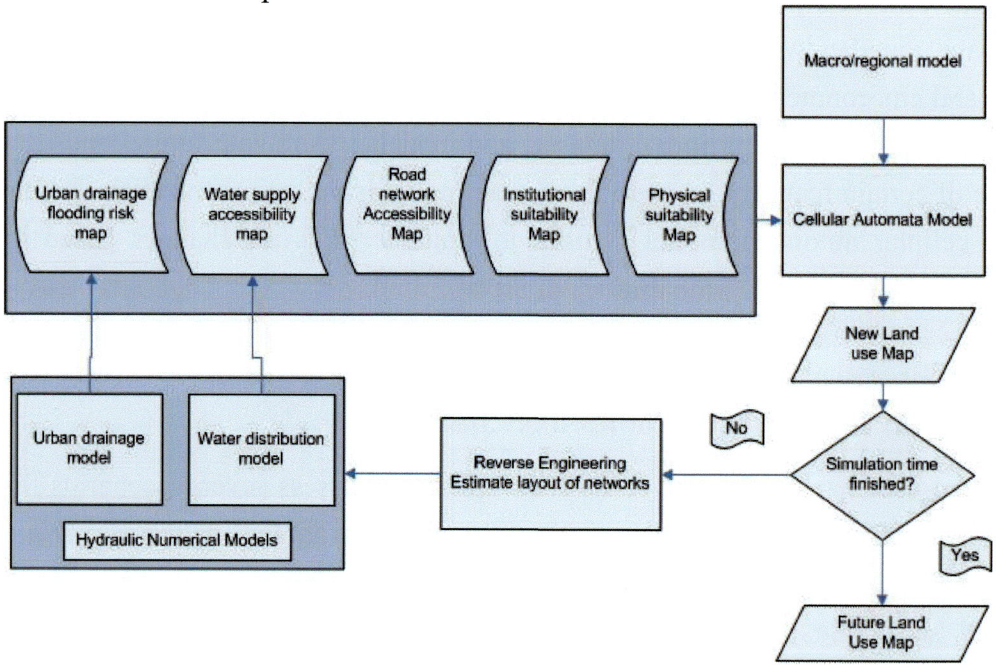

Figure 6.1 A simulation loop within the framework

The work described here explores the integration of a cellular automata model with information coming from urban water models to assess the suitability of urbanization. There is a relation between urban development and its impact to the existing water infrastructure that affect the availability of water services and therefore the suitability for urbanization. At the same time new infrastructure will be developed as a result of the urbanization process that again affects the suitability for urban land. The integration of these models allows the exploration of several planning scenarios to assess the impact of certain actions, policies, regulations and to explore several urban futures.

For urban drainage in particular the connection of cellular automata models for land-use change with spatial data analysis and urban drainage network models is showing promising results. The research has shown that by analyzing the spatial relation between the drainage network, the road network and the land-use, knowledge about the positioning of the main drain conduits can be derived. This yield to a new approach to derive the layout of drainage networks of existing systems that can be used to asses scenarios of investments and rehabilitation. Moreover, the approach can be applied to develop case studies in any city in the planet with information currently available on the Internet. Any derived case study can be optimized against performance objectives and therefore be compared to the existing system. This approach can conduct to a new indicator that can provide decision makers with information about how sub-optimal the existing system is and the level on investment required to upgrade the capacity.

The proposed approach also has been tested to build future scenarios of urban development that conduct to the possibility of deriving a future network layout. This was optimized to be sized and therefore connected to the existing model. This allows the evaluation of the impact of the future developments on the existing network. By doing this, critical elements of the network can be identified and rehabilitation strategies can be tested in advance.

There are still limitations of this method, the type of data required, the quality of the data and the availability of data affect the application of this approach. More cases need to be developed for further analysis. Information and data from climate change scenarios can be included to expand the capabilities of the analysis.

The connection of urban drainage model with the cellular automata of land-use change can be further explored by proposing an objective function that relates the area and type of future developments in a city to an index that includes the future implications or challenges to manage the urban drainage system. This can lead to a new set of tools and techniques to support urban planning but also guide decision makers not only on when to increase the population density in parts of a city but also where. Furthermore, this line of research can be used to describe the impact of the provision of water services as a trigger to stimulate urban development. The results will lead to scientific publication but also to practical tools and approaches that can be used by several municipalities to enhance urban planning.

The expected result is a new approach for urban water infrastructure planning which help water companies and municipalities to have a dynamic planning tool to improve the effectiveness of their investments and to be more environmentally efficient.

6.4 Application to the Case Study

6.4.1 Background

Belo Horizonte (BH)is the capital of the State of Minas Gerais, which in economic terms (gross product) is the third among the 26 Brazilian states. The city lies at 20° South latitude and 44° West longitude (Figure 6.2) and has an altitude of 750 to 1,300 meters. It is located in a mountainous region of tropical soils that originated from the decomposition of metamorphic rock. Tropical highlands weather predominates in this area, with average yearly rainfall of 1,500 mm and average yearly temperature of 26°C. The rainy season lasts from October to March, when 90% of the total yearly rainfall occurs. The highest monthly average rainfall (315mm) takes place in December. Typical rainfall intensities are also relatively high (e.g.: 200 mm/h in the case of a 10-year return period event with 5 minutes duration; 70 mm/h for the 1h and 50-year return period event). Mean relative humidity reaches 50% during winter and 75% in summer.

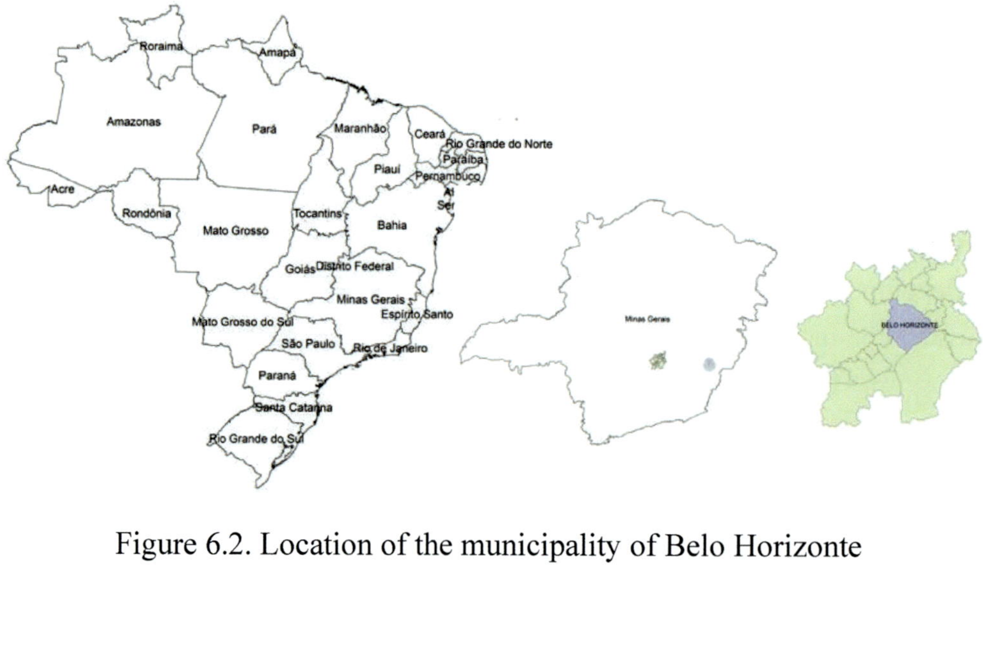

Figure 6.2. Location of the municipality of Belo Horizonte

Belo Horizonte has 2,227,400 inhabitants with a population density of 6,900 inhabitants/km^2. It is a planned city, built in 1898 to become the capital of the state. The total area of the municipality is 330 km^2. Belo Horizonte is inserted into a metropolitan area; the RMBH (Belo Horizonte Metropolitan Area), gathering 33 distinct municipalities, with an area of 9, 179 km^2 and 3,900,000 inhabitants.

The Belo Horizonte territory locates at two main catchments (Arrudas creek and Onca creek catchments), each representing at about 50% of the total municipal area. Part of those catchments locates at neighbourhood municipalities: Contagem, upstream of Belo Horizonte, and Sabará and Santa Luzia, downstream of Belo Horizonte. There are no rivers in the municipal territory, although Arrudas and Onça are direct tributaries of the Velhas River, with a total drainage area of about 40,000 km2, which itself is the tributary of the Sao Francisco River, the longest one entirely within Brazilian territory (approximately 600,000 km^2 of drainage area).

Stormwater management has been entirely under the responsibility of the Belo Horizonte municipality since the city foundation. Traditional storm water systems prevail in the city, although experiences with detention ponds exist since the 50s. There are at about 4,300 km of roads all of them equipped with gutters, inlets etc. The municipal database on drainage infrastructure keeps details about 64,000 inlets (gullies), 11,500 manholes, 1,100 outflow structures (outfalls), and almost 770 km of stormwater sewers. There are some 700 km of perennial creeks in the municipal area. Part of those creeks have been lined, most of them as culvert concrete channels. The length of lined channels reaches near to 200 km (Figures 6.3).

The creek lining policy, which prevailed up to the 90s, was mainly justified by the following rationale:

- Lining is required for increasing the flow velocity and the channel conveyance, reducing the flood risk;
- Lining makes easy the implantation of interceptor pipelines and the so called sanitary roads;
- Lining makes easy the creek maintenance;
- Health risk due to directed contact with polluted waters may be reduced by creek lining;
- Inhabitants of riparian zones usually ask for creeks to be lined

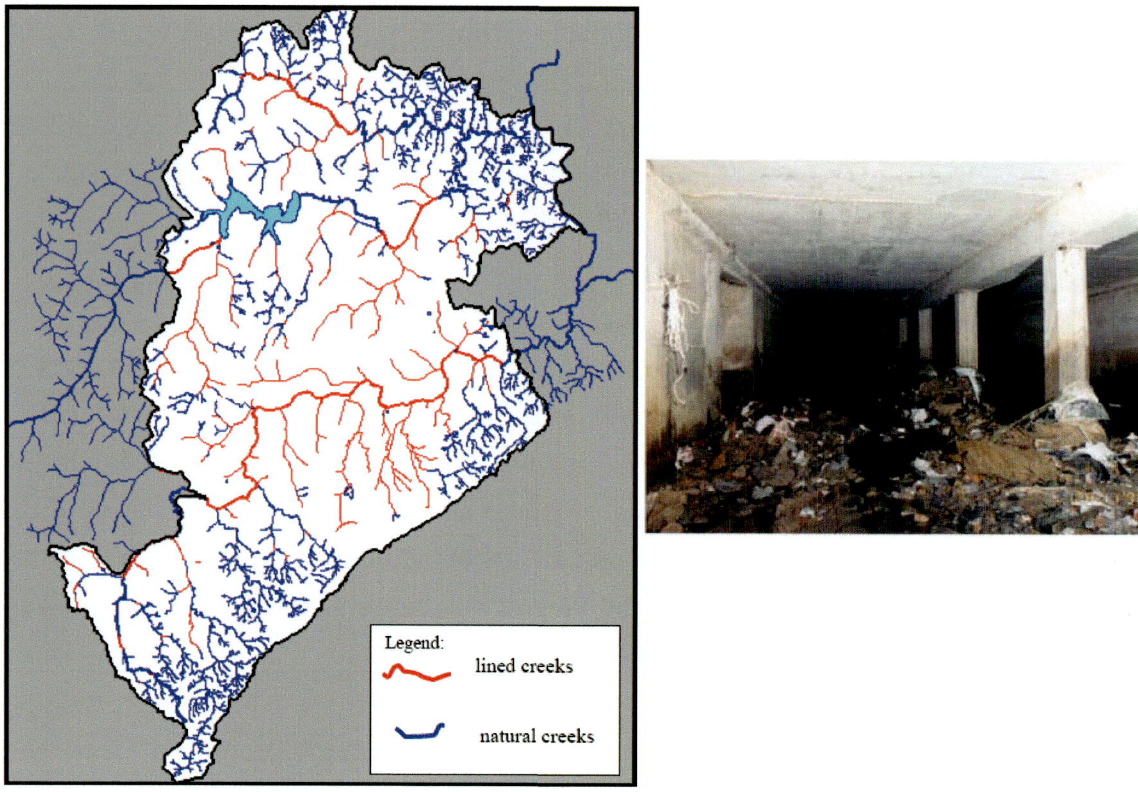

Figure 6.3 Belo Horizonte hydrography and lined channel photo

The apparent simplicity of stormwater management, as perceived almost during all the last century, led to the use of very simple design methods for storm water systems. Synthetic models were used which do not require observed data to calibrate parameters (e.g.: rational method and synthetic unit hydrograph). Since observed data were considered as not necessary for storm water management, during all the last century the Belo Horizonte municipality did not invest in monitoring stream discharges or water quality parameters. One of the consequences of those approaches is high uncertainty in hydrologic design. A similar oversimplification also prevailed in hydraulic design. Complex flow conditions, including the effects of stream confluence, flow transitions or unsteady flow, were infrequently regarded and model simulations of these conditions were rarely done. Only uniform flow conditions were used to assess if the pipe network needed to be increased.

6.4.2 Data collection

For this case the land-use map for the year 2007 was collected in the framework of the learning alliance in Belo Horizonte. Maps for other years exist but were not available for this research. Because the focus of the study is the urban areas a higher level of detail is considered for artificial areas, therefore, the land-use classes were grouped into 6 classes mainly grouping the parks, lakes, airports into not modelled class. The area outside the administrative perimeter of Belo Horizonte was considered vacant land, this is assumed since there are other urban agglomeration in the vicinity of Belo, but it will help to see how the city may expand if all that land is available for growth. The other classes that are considered are the residential areas with high, middle and low income. The industrial, commercial and institutional areas are mixed in one class. Figure 6.4 shows the land-use map.

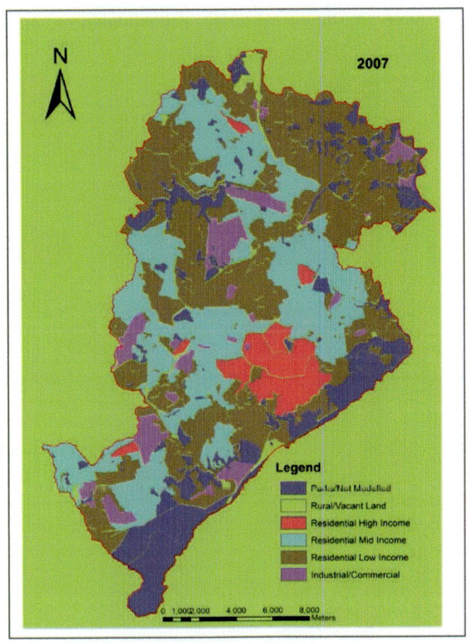

Figure 6.4 Land-use map for Belo Horizonte for year 2007

A spatial dataset was collected for the study area. The dataset consisted of following maps: elevation (DTM), derived slope map, main trunks and motorways, primary and secondary roads, rail network, and the shape files with the lines of the drainage network and the water distribution network. For the Digital Terrain Model (DTM) the ASTER Global Digital Elevation Model (ASTER GDEM), data with 30 meters resolution, was used. The DTM were re-projected and clipped into the proper extension of the study area. The DTM was used to generate the slope raster map.

The shape file representing the motorways, trunks, primary and secondary roads, rails, were downloaded from open street map project. (http://www.openstreetmap.org/). Shape files with the lines of the urban drainage and water distribution networks covering the area were made available through the learning alliance established in Belo Horizonte. For the study area there were not models available for the drainage and water distribution networks. Moreover, the available spatial information did not include information related to the geometry of the drainage and water distribution networks. As was presented and discussed in chapters 4 and 5 the information of the geometry of the elements of the network is important to classify the information and find spatial relations. Despite this limitation, the analysis was conducted with this dataset. The maps are presented in figure 6.5

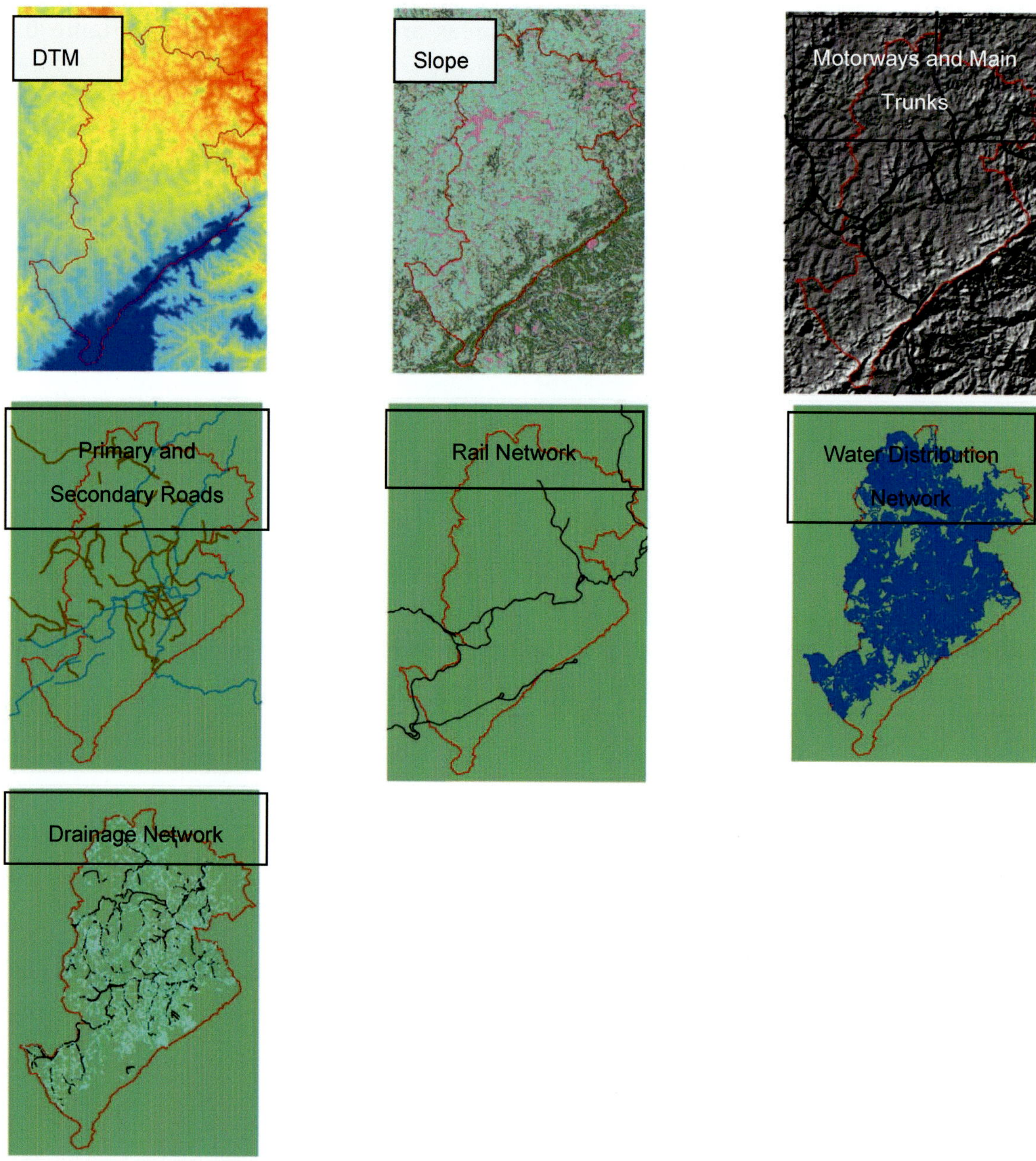

Figure 6.5 Spatial dataset used for Belo Horizonte.

6.4.3 Land-use and urban drainage system

Following the method describe in chapter 5 a corridor analysis was done along the main drainage pipes to derive a connection between the land-use and the existing urban drainage network. For this purpose the information concerning the lines that represent the interceptors was used for this procedure. Figure 6.6 Left. shows the result of the analysis for the land-uses: 'residential high income', 'industrial/commercial', 'parks' and 'recreational'. Figure 6.6 right shows the result for the land-use classes residential middle and low income.

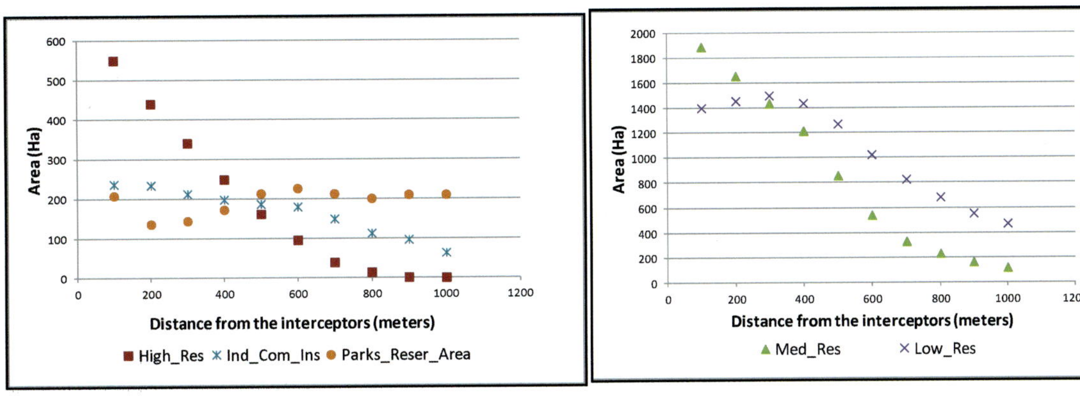

Figure 6.6. Land-use area distribution with distance from the main interceptors

The information obtained from the corridor analysis shows a similar trend as the ones obtained for the case of Birmingham in the UK for drainage networks. For the case of Belo Horizonte the land-use classes high and middle income residential areas show a similar trend, the intercepted area is declining with the distance from the main interceptors until a distance of approximately 800 meters. For the land-use classes low income residential and industrial/commercial/institutional the trend shows a break point at approximately 500 meters. Like in the case of Birmingham this information can be used to anticipate the location of new pipes in futures scenarios based on the estimated land-use distribution. This means that the main collectors of the network are positioned within 500 meters of these land-use categories. It is possible that this finding reflects the traditional engineering practice of using a higher return period in the design and construction of drainage networks inside urban areas to provide a higher degree of safety to valuable parts of the city including governmental institutions, hospitals, industry, etc. The presence of the class low income for the case of Belo Horizonte can be possible explained tend to locate in the vicinity of rivers or areas that are pruned to flooding due to the cost of the land. It is possible that the interceptor were constructed at a later stage when those urban developments were already there, but that information is missing at this point in the research to be sure of this argument.

Another important aspect of urban drainage is that the drainage is largely determined by the topography. In fact, this is the main criterion used for the design of urban drainage networks. Figure 6.7 presents a comparison between the existing main interceptor network in Belo Horizonte and the derived natural drainage network obtained by processing the DTM. The flow accumulation map was used to derive the natural drainage.

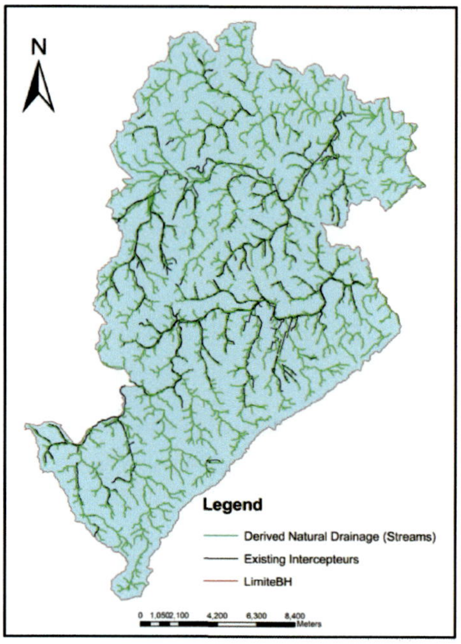

Figure 6.7 Existing urban drainage network and derived natural drainage network, Belo Horizonte

Figure 6.7 shows a relatively high spatial correlation between the constructed drainage network and the GIS-derived natural drainage network produced for the terrain topography. The spatial analysis shows that it is possible to relate information from the land-use map and the topography (elevation and/or slope) with the existing drainage infrastructure.

6.4.4 Deriving the network layout for the existing system

To apply the proposed approach presented and discussed in chapter 5 the centroids of the land-use clusters of interest found with the corridor analysis were calculated. The centroids correspond to the land-uses residential high, middle and low income, as well as the industrial/commercial.

The land-use information, topography, slope and the road network, was combined to create a cost-weighted raster to generate the drainage network layout. The cost-weighted raster is a map that contains a cost factor for each cell. With this map it is possible to identify the least cost path (the one with the minimum total cost) from the starting point towards the outlet of the catchment or the exit manhole. The exit manhole in this case was assumed to be the location of the main wastewater treatment plants in Belo Horizonte.

The generation of the drainage network layout is achieved by crossing areas with gentle slopes, following the alignment of the existing roads (primary and secondary roads) and passing through 'vacant land', 'residential high income' and 'industrial/commercial' land-uses. The outcome of the simulation applied to the municipality area of Belo Horizonte is shown in Figure 6.8 and figure 6.9. The layout was derived for different combinations of land-use centroids of interest to assess the effect of the land-use classes in the generated layout. Figure 6.6 shows the simulation for the land-uses classes 1 (residential high income), 2 (residential middle income) and 4 (industrial/commercial).

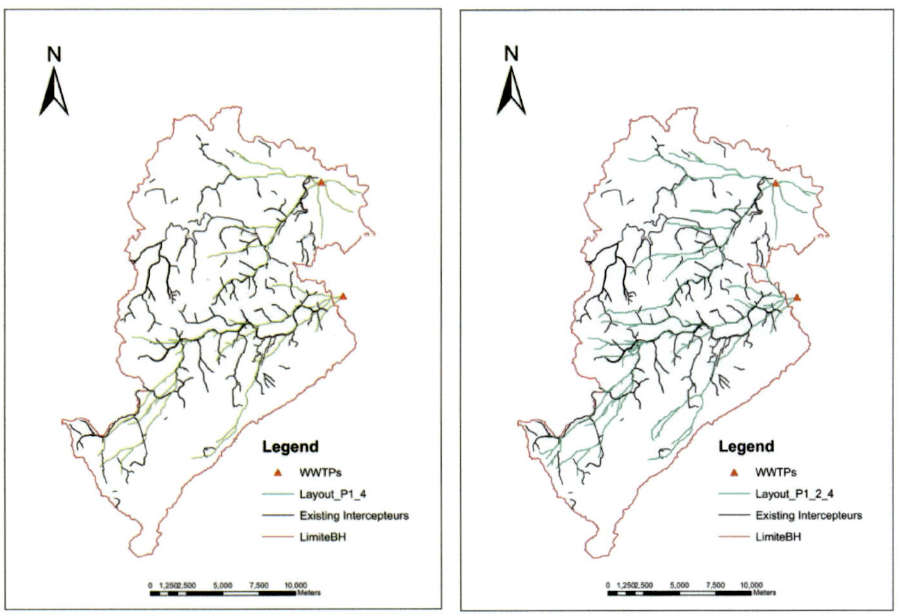

Figure 6.8. Drainage network layout generated using land-use classes 1 and 4 (left) and land-use classes 1, 2 and 4 (right).

The results of the simulation for the land-uses classes 2 (residential middle income), 3 (residential low income) and 4 (industrial/commercial) are shown in figure 6.9.

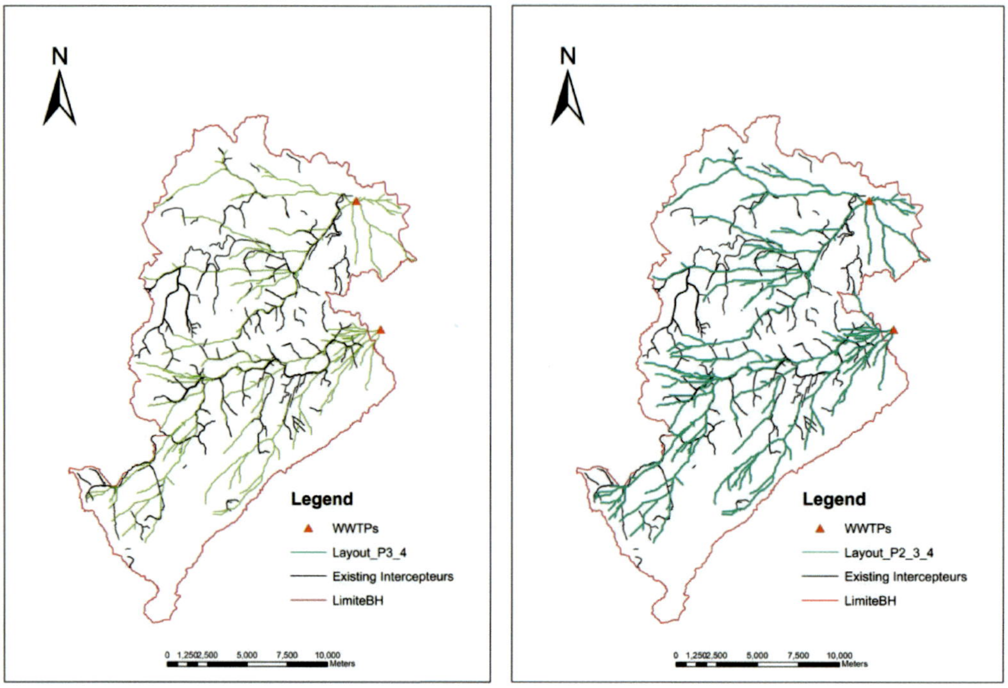

Figure 6.9. Drainage network layout generated using land-use classes 3 and 4 (left) and land-use classes 2, 3 and 4 (right).

The results presented in figures 6.8 and 6.9 shows that the obtained layout approximates well to the layout formed by the main interceptor lines in the Belo Horizonte municipality. The Hausdorff method was used as a measure of fitness as indicated in the methodology. The results obtained are summarized in Table 6.1.

Table 6.1: Hausdorff distance for the generated sewer layout.

Land-use classes	**Cost Weighted Raster**
1 and 4.	362
1, 2 and 4	383
3 and 4	332
2, 3 and 4	361

Figures 6.8 and 6.9 together with Table 6.1 show that the average separation between the existing interceptor network and the simulated layout is between 332 - 383 meters. For all the tested combinations the Hausdorff distance is within the same order of magnitude and the biggest difference will be around 50 meters. The best approximation is achieved when the centroids of the land-use clusters 3 and 4 are used. These two land classes correspond to residential low income and industrial/commercial classes, which were the more attractive land-uses found with the corridor analysis. The results

obtained with the application of the method to the Belo Horizonte municipality area are similar of those obtained for the case study of Birmingham and presented in chapter 5. The Hausdorff distance obtained with the simulated layouts for the drainage network in Birmingham was around 150 meters, which is less than half of that obtained for Belo Horizonte. This is possible due to the use of a better data set for both the drainage network and land-use maps. The spatial coverage of the simulated layout obtained with the land-use classes 3 and 4 is good, when compared with the others. Confirming that the land-uses 1 and 2 do not provide additional "knowledge" for the spatial position of the main drainage pipes. This link is important to use this approach to test future scenarios of urban land-use changes.

6.4.5 Modelling land-use change

The approach described in the previous section is empirically tested in the area of the municipality of Belo Horizonte in Brazil. This is done in order to observe the land-use dynamics in the study area and to apply the concepts developed in chapters 3 and chapter5 in particular.

Since the land-use dataset was available for only one year, the simulation is defined as an empirical scenario. There are no other land-uses maps to statistically compute the weights of evidence of the different variables. Therefore, these are assumed based on the findings of chapter 3 and some general rules of attraction-repulsion such as the one described by Hagoort, 2006.

The aim of this simulation is to probe a concept rather than predict the exact locations of particular land-use developments. For this reason the interest is to observe the general spatial structure that emerges in the case the city of Belo Horizonte expands towards its vicinity and if all that land is available for this organic growth. The result of the simulation is assessed visually by comparing the simulated land-use map to an existing previous study conducted in this area.

Land-uses and study area

The modelling area includes the administrative area of Belo Horizonte and neighbouring municipalities. The area outside Belo Horizonte was considered vacant land, this is assumed since there are other urban agglomeration in the vicinity of Belo, but it will help to see how the city may expand if all that land is available for growth.

The classes that are modelled are the residential areas with high, middle and low income. The other class includes the industrial, commercial and institutional areas which are mixed in one class. The land-uses parks, lakes; airports were grouped into class zero or not modelled area within the study area.

The spatial resolution is chosen such that cell size allows for the capture of the differences and peculiarities of the intra-urban space at the block level. Cell sizes in applications of urban models usually vary between 50 and 150 meters. The cell resolution is defined as 30 by 30 meters, equivalent to an area of 0.09 Ha. The spatial data set described in section 6.1.1 is used to build the land-use change model in Dinamica EGO.

The study area is presented in figure 6.10 and its position with the neighbouring municipalities of Belo Horizonte. The area contains the jurisdiction of Belo Horizonte, Barreiro and Vendanova. The areas considered as vacant land in the simulation contains areas of the neighbouring municipalities of Contagem , Ibirite, Nova Lima, Ribeirao das Neves, Justinapolis, Sabara, Santa Luzia and Vespasiano.

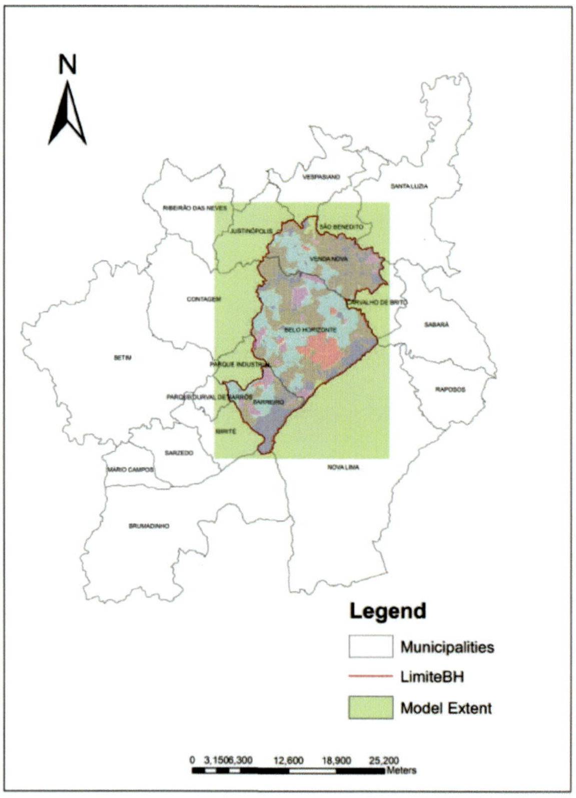

Figure 6.10 Location of the modelled area

The lattice or grid of the workspace contains 795 rows and 1122 columns, which are equivalent to 891,990 cells. This contains more than 500,000 cells that are considered as vacant land.

Transition matrix

The transition matrix considers the demand for land (cells) of each class. For this exercise dataset for different years were not available. Therefore, the demands are based in the study done by Furtado, 2009 in the area of Belo Horizonte. Table 6.2 presents the distribution of residential land-uses in Belo Horizonte described in Furtado, 2009.

Table 6.2 Number of cells occupied for residential land-uses.

Year	1991		2000		Difference
Land-use class	Number of cells	Percentage (%)	Number of cells	Percentage (%)	Number of cells
High-income	3321	0.04	5850	0.063	2529
Average-income	27164	0.36	31672	0.340	4508
Low-income	44198	0.59	55570	0.597	11372
Total	74683	100	93092	100	n/a

Source: modified from Furtado 2009

During the decade of 1991 to 2000 the number of inhabitants increases 22%, from 3,212,044 in 1991 to 3,904,172 in 2000, according with the data presented by Furtado 2009. To construct the transition matrix for this simulation it is assumed that the same level of growth occurred during the decade 1991-2000 will occur in the next 30 years. This is because the population annual growth for the state of Minas Gerais in Brazil is expected to be less than 1% in the coming years. Table 6.3 presents the transition rates estimated to build the simulation for the year 2037.

Table 6.3 Estimated transition rates for the year 2037

Land us class	Number of cells	Transition rate for 30 years.	Average transition rate for 10 years.
High-income	22761	0.044	0.0148
Average-income	40572	0.079	0.0264
Low-income	102348	0.199	0.0665
Industrial/Commercial	26655	0.052	0.0173
Available land	512476		

The number of cells presented in table 6.3 corresponds to the difference between the years 1991 and 2000 presented in table 6.2 multiply by 9 to consider the change in cell resolution from almost 90 by 90 in the study of Furtado, 2009 to 30 by 30 in this case.

The proportions of residential land-uses are kept equal as those in the 90's that mean according with the land-use classification that the distribution of the population and the income will not change significantly. To estimate the demand for commercial and industrial areas, the current distribution of land-uses was considered. In this case the proportion of cells for commercial and industrial uses is about 1% of the occupied land.

Weights of Evidence and CA Rules

To simulate the land-use changes in Belo Horizonte, the weights of evidence obtained in the case of Birmingham and Villavicencio were used. This was done due to the lack of data to compute the weights of evidence for Belo itself. It is recognized that there will be regional and /or specific differences in the estimation of this weights, but this method is the heart of land-use dynamics used by Dinamica EGO. For this reasons, the estimated parameters for Birmingham and Villavicencio were used as a skeleton and were further adjusted for Belo Horizonte. The shape of the attraction/repulsion figures for the drainage network was reviewed based on the corridor analysis done for Belo Horizonte. For the case of the water distribution network the shape of the attraction/repulsion figures were taken similar to those found in the corridor analysis for the case of Villavicencio. Furthermore, the weights of evidence found for Birmingham and Villavicencio were approximated to follow the "six general rule shapes" proposed by Hagoort, 2006 and Geertman et al 2007. According with these authors and based on an extensive research conducted in the Netherlands, neighbourhood effects might have the following general behaviour (see figure 6.11):

1) (I and II) a decreasing net positive (or negative) relation that becomes neutral with distance;
2) (III and IV) a net negative (or positive) relationship that switches to a net positive (or negative) relationship and then becomes neutral; and
3) (V and VI) an increasing net positive (or negative) relation that switches to a decreasing relationship, and becomes neutral.

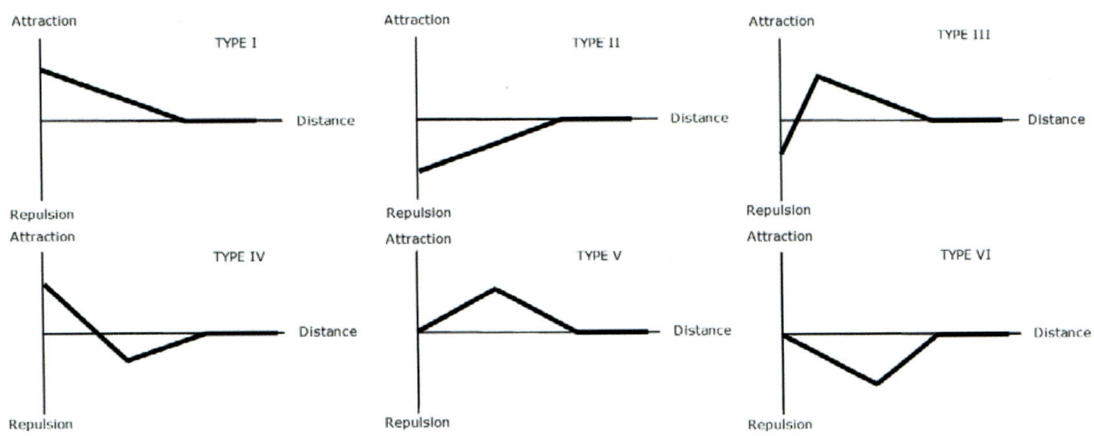

Figure 6.11. Six general rule shapes to model land-uses changes (Hagoort 2006, Geertmant et al 2007)

For example in figure 6.12 it is presented the weights of evidence figure for the transition from vacant land to residential low density (low income for the case of Belo Horizonte) the importance (weight) of distance from the land-use industrial/commercial.

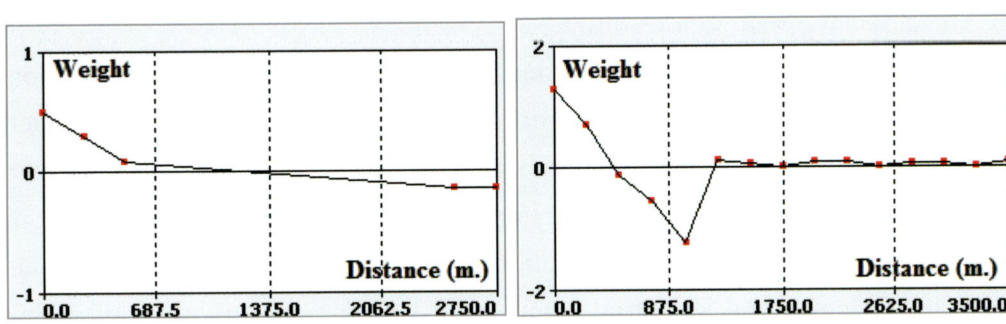

Figure 6.12 Example of adaptation of Weight of Evidence (attraction/repulsion) for the transition from vacant land to residential low income distance to industrial/commercial areas.

Figure 6.12 shows that for the case of Birmingham the Weights of evidence figure approximates well to the rule type I of Hagoort presented in figure 6.11 or general rule shapes. For the case of Belo Horizonte this rule was transformed to the type IV of Hagoort reflecting that for low income class being close to industrial areas may be attractive at short distances depending on the type of industry and become repulsive after some distance to later become neutral.

Furtado, 2009 followed the approach of the six general rule shapes to model the land-use changes in Belo Horizonte for the year 2050. According with the approach followed in his study for Belo Horizonte the application of the six general rule shapes was as follows:

Rule type I proposed by Hagoort (2006) applies to the impact of residential classes on attractiveness for themselves and the impacts of 'small service and commerce', 'large service and commerce', and 'institutional buildings' on residential use – that is, they all have decreasing net positive relationships.

Regarding attractiveness between actors, both high-income and average-income actors attract each other (type I). However, average-income actors are attracted much strongly than high-income actors. Proximity to low-income actors has a negative impact on attractiveness for high-income actors (type II).

The approximation of the previously obtained weights of evidence attraction/repulsion figures to the six general rule shapes was done manually in this particular case for all the transitions and all the variables.

Model Parameters

The parameters for the functions expander and patcher were assumed based on the results obtained for the case study in Birmingham presented in chapter 3 after the calibration. The parameters mean patch size and variance of the patch size were assumed in general as 2 Ha. and 150 Ha. respectively for the expander function. Similar values were used for the patcher formation with small difference in particular for residential high income and industrial/commercial land-uses were values of 3 Ha. and 130 Ha. were used. The patch isometry parameter was increased from 1 used for Birmingham to 1.5 - 1.7 in the case of Belo, this helps in achieving a more symmetric shape of the cluster. The parameters for the function that regulates the transition done by the expander or the patcher were set to 0.97 - 0.99, so that most of the transitions are carried out by the expander function.

Land-use for year 2037

With all the parameters in place the model was run. The final map obtained after several runs to adjust some parameters with the objective of achieving a more symmetric shape in some clusters is presented in figure 6.13. Since we do not have a final map to compare the outcome of the simulation for the year 2037 the inspection was done visually. The outcome of one the simulation performed by Furtado 2009 using the Moland framework is shown in figure 6.14 it was used for visual comparison with the one obtained using Dinamica EGO.

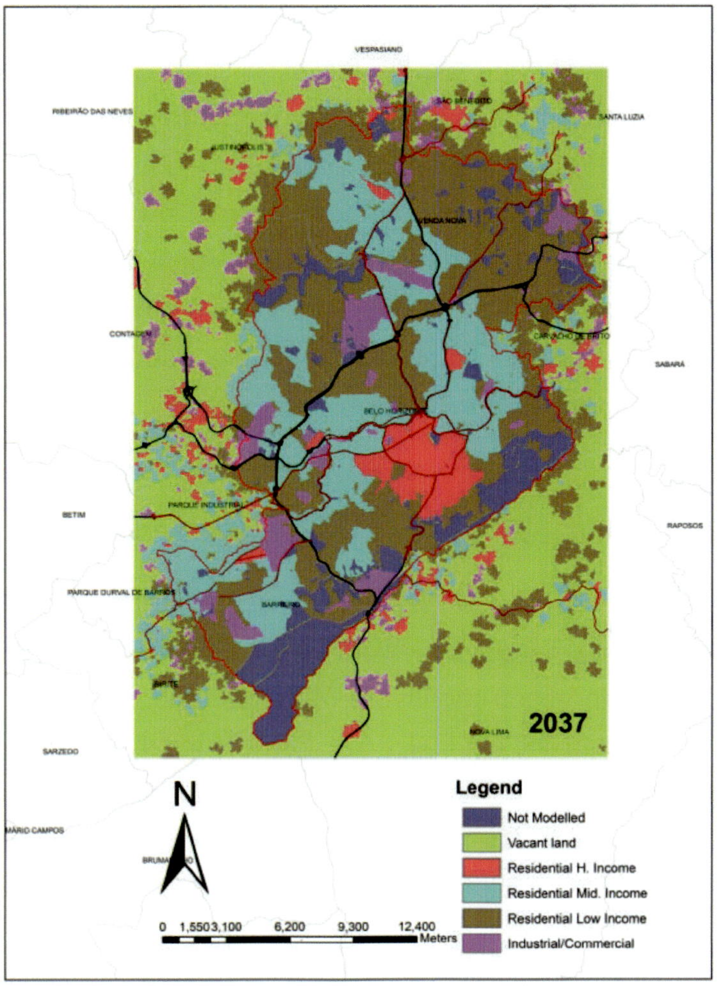

Figure 6.13 Simulated land-use map of Belo Horizonte for year 2037

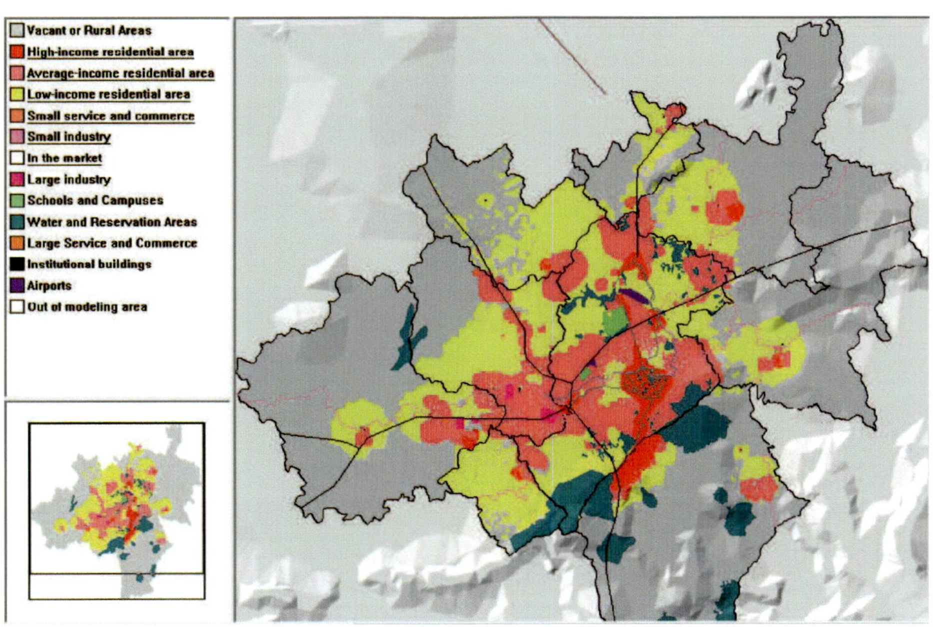

Figure 6.14 Land-use map of Belo Horizonte for year 2050. Furtado, 2009

A visual comparison between figures 6.13 and 6.14 shows that despite the fact that they are generated with different simulation engines and methods and for different years, there are some similarities. For example the expansion of the low income and mid income residential classes in the south west of the study area. The expansion of the middle income class in the mid west of the study area is also similar. There is also a new cluster established in the north part of the study area towards the centre of the Santa Luzia area. The application of the modified local rules obtained with Dinamica Ego for the previous case studies to follow the six general rules proposed for Hagoort and others to model urban land use changes can reproduce a similar land use pattern for the case of Belo Horizonte for the formulated scenario.

6.4.6 Deriving the Network layout for the year 2037

The simulated land use map for the year 2037 presented in figure 6.13 was used as the input to generate the future drainage network layout. To do this the same rules used in the cost weighted rates to obtain the layout of the present system shown in figure 6.9 were used. The centroids of the land uses categories 3 and 4 were obtained and used as the starting point. Figure 6.15 shows the result of the simulation for the year 2037. The network was generated towards the position of the wastewater treatment plant.

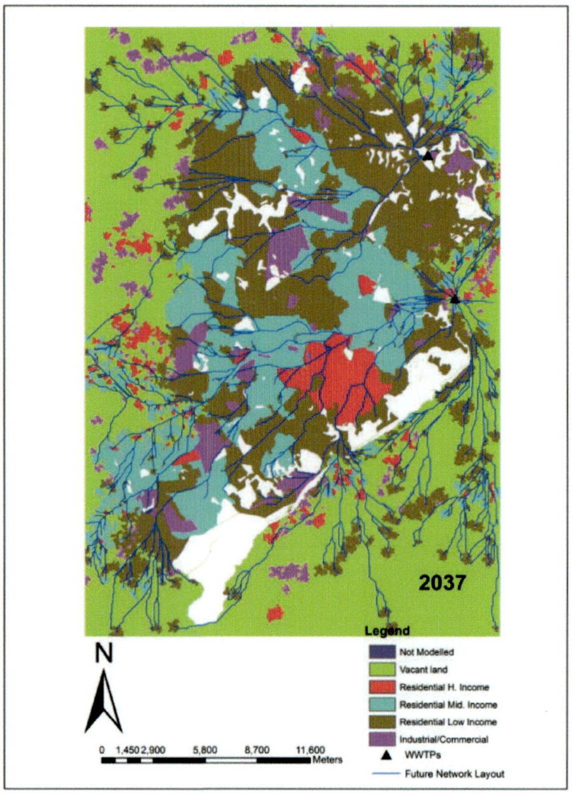

Figure 6.15 Future drainage network layout

Figure 6.15 shows a possibility for the future drainage network layout for the Belo Horizonte area. This layout was obtained considering that the system in the future will interconnect the new areas of development. Nevertheless, care must be taken when exploring the future technological option to assess if that is physically possible. For instance, it is still possible that some new areas of development will not occur in the same catchment and therefore there is no gradient to do such interconnection. To illustrate this, a drainage network layout was generated by assuming that the future developments will be connected to the closest existing drainage pipe, for economic reasons. To do this, the same rules were applied as the ones used to derived figure 6.15 but the flow will not be directed to the wastewater treatment plant but to the drainage layout presented in figure 6.9. the result is shown in figure 6.16.

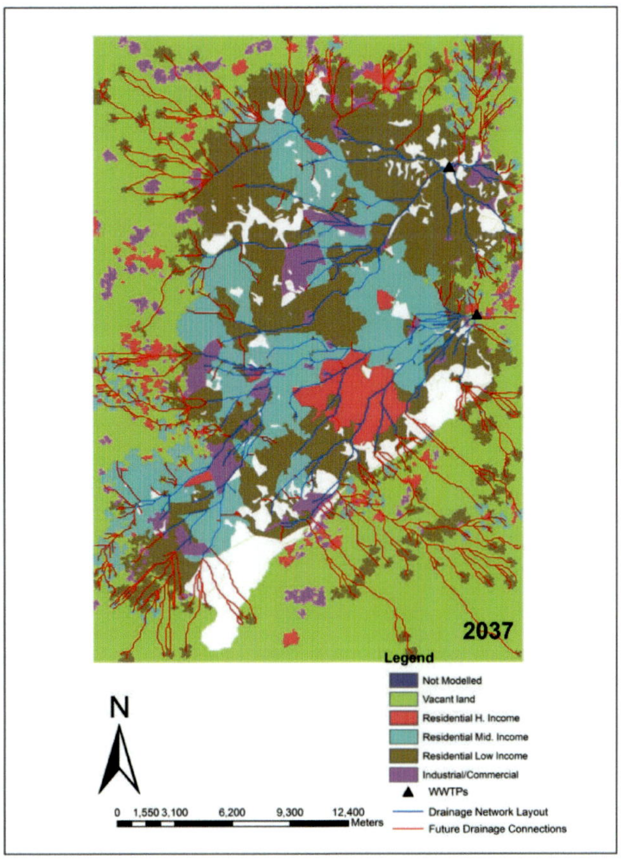

Figure 6.16 Future drainage network layout interconnected to the existing pipe system.

The results of figure 6.16 in comparison with figure 6.15 shows a more directed connection towards the existing pipe system, the lines are more straight. This is clearly observable in the south east part of the system, but that connection will be difficult in reality since that area correspond to a mountainous area and the flow will be uphill.

6.4.7 Drainage model

Within the framework of the SWITCH project a portion of Belo Horizonte's drainage network was setup in Mouse to be tested. The topography and the information to build the model were provided by the local partners in the learning alliance. The information was Pre-processed in ArcGis 9.3 to define the sub-catchments and main streams. This part of the network corresponds to the Vendanova Catchment located in the North side of Belo, as is shown in figure 6.17.

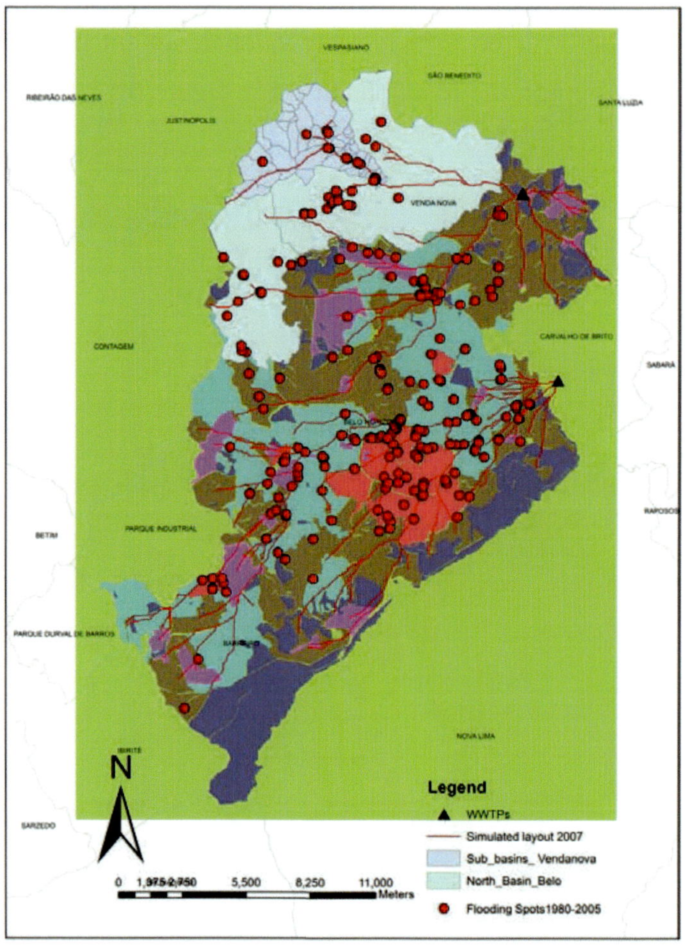

Figure 6.17 Location of the Vendanova cachment.

The network is composed by 168 pipes and 169 nodes see Figure 6.18. A rainfall event corresponding to a precipitation of 20 mm over a period of 6 hours was used for the simulation. Figure 6.18 shows the network schematization developed in SWMM for this case study and the hydraulic profile during the peak flow. The flooding nodes are shown in the schematization and their position in the network are similar to those reported previously in flooding events during 1980-2005 reported by the learning alliance in Belo Horizonte.

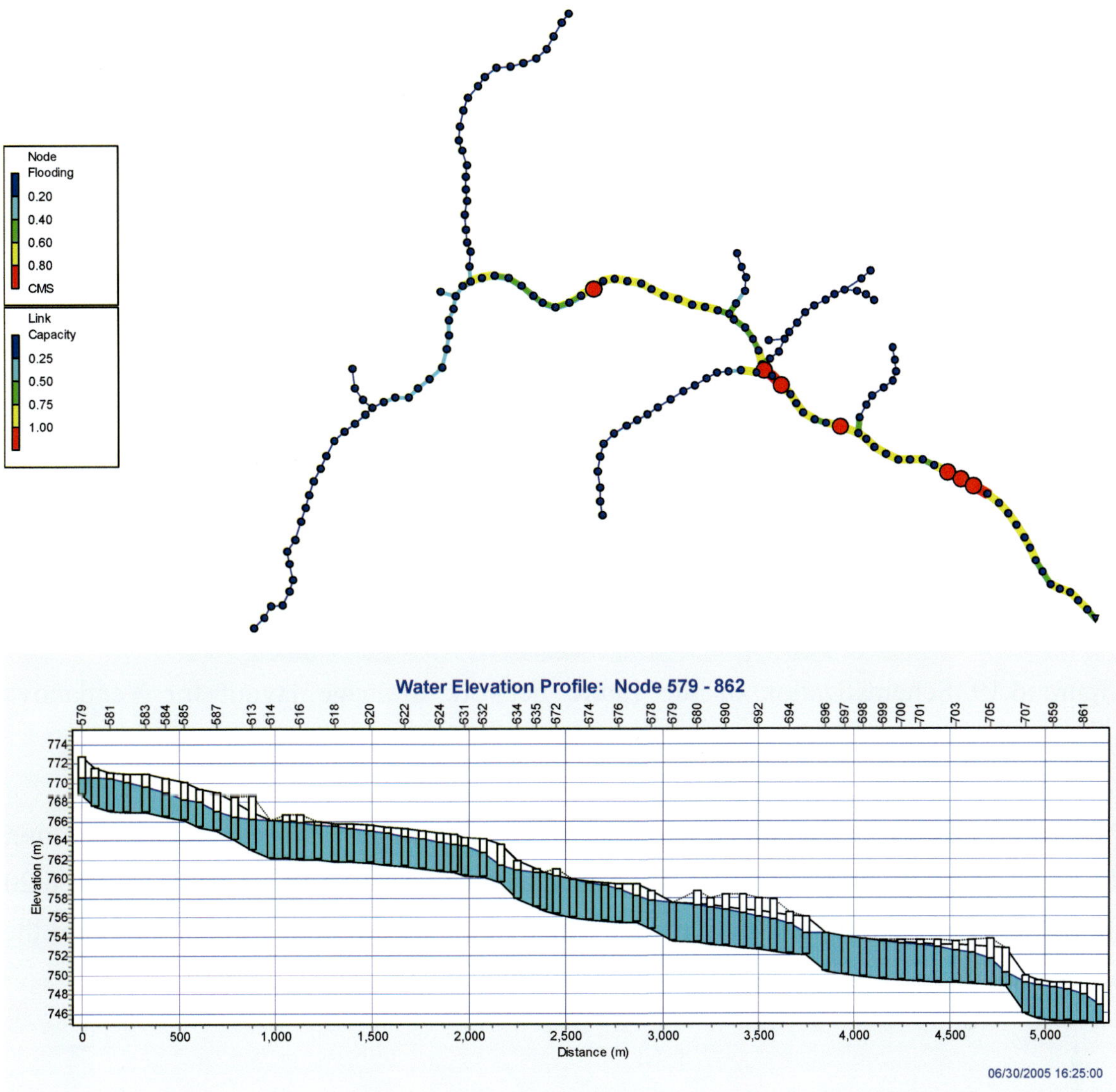

Figure 6.18 Model schematization in swmm, Plan view with flooding node and hydraulic profile.

Figure 6.18 shows that for this rainfall event all the nodes that are flooded occur in the central collector of the system and the location of the flooding nodes correspond well with those reported in reality and shown in figure 6.17 for this catchment. In the framework of SWITCH this network was optimized and the optimization process was setup to solve the flooding situation in the system with two objective functions, Flooding damage and pipe renewal cost. More details on the optimization process that was followed can be found in SWITCH, 2010 and Barreto 2012.

To assess the performance of the network for the future scenario of land use change for the year 2037, the derived layout presented in figure 6.15 was processed to connect the future developments around this catchment with this part of the drainage network. Figure 6.19 shows the schematization of this part of the network.

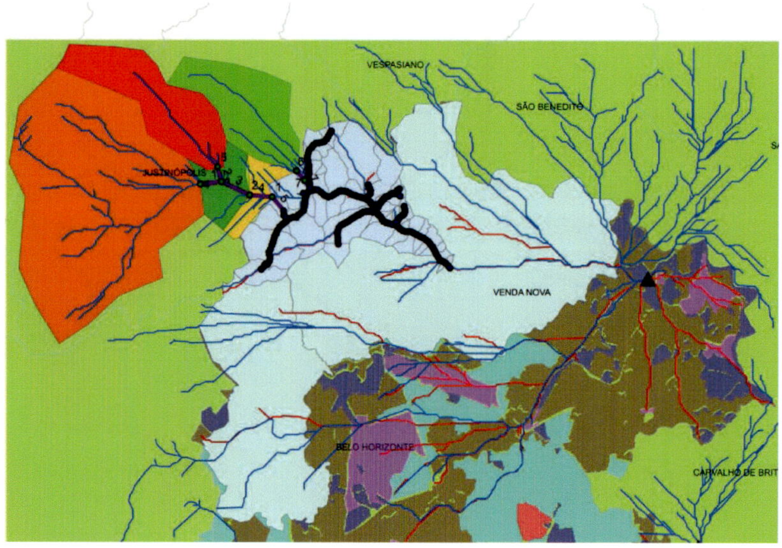

Figure 6.19 Schematization of the future network drainage layout for Vendanova catchment.

According with the derived drainage network layout for the future scenario of land use, the parameters to setup the model were estimated and exported to swmm. Figure 6.20 shows the plan view of the swmm model. The hydraulic profile for the connection of the new development with the existing model of Vendanova is shown in figure 6.21.

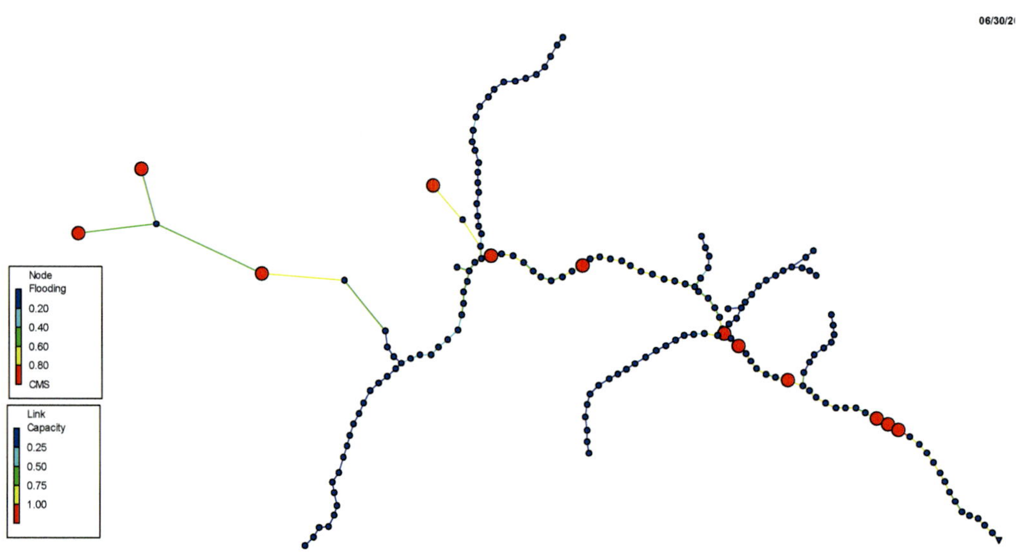

Figure 6.20 Drainage network model for future scenario in Vendanova

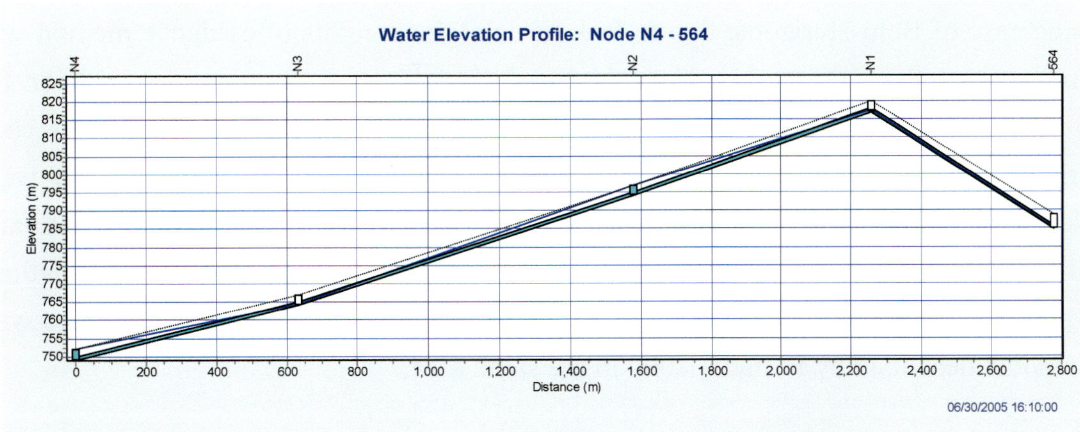

Figure 6.21 Hydraulic profile new connection for scenario 2037 in Vendanova.

This connection to the existing network is used here as an example, to verify the outcome of the model and check if the result obtained with the simulation are possible to be implemented in reality. In this case, even though the outcome of the layout generator look good visually it is important to check the direction of the flow and the topography. This also highlights the need for further improvement of the tools developed here; particularly important to generate a simulation platform that allows the integration of land use change models with physics-based model of water networks. This also shows the limitations of any automatic, semi automatic modelling technique and the need of engineering criteria, knowledge and expertise.

6.5 Conclusion

The integration of cellular automata models of urban land use changes with physics-based models of water networks is explored in this research. This chapter aimed to generalize the approach and it was tested in a different case study. For developing countries the introduction of this approach may not be easy, it requires the compilation of several spatial datasets and the existence of models, tools, but also capabilities. In the case of developed countries the information and data may be available, but processing the information is still a demanding task. The approach can be computationally demanding, in particular the optimization components. However, the application of the proposed approach to the study area of Belo Horizonte, with limitation in the information and data, demonstrates that this approach is valuable for testing scenarios of urban development and water infrastructure planning.

In the case of Belo Horizonte due to lack of data, the weights of evidence method was combine with the six general rule method to derive the rules of attraction/repulsion for the CA engine. This process was done manually manipulating the weights of each variable and transition. The parameters for the functions expander and patcher were selected according with the calibration datasets found for the case of Birmingham and Villavicencio. The results, even thought, are not intended to determine exact predictions of particular land use classes but more as a demonstrator, show several similarities with the model described by Furtado, 2009 in the same area.

The application of the proposed approach in Belo Horizonte has shown that by analyzing the spatial relation between the drainage network, the road network and the land-use, knowledge about the positioning of the main drain conduits can be derived, similar to what was found in Birmingham. This approach can be used to derive the layout of drainage networks of existing systems that can be used to asses scenarios of investments and rehabilitation, particularly important in cities that have growth without proper drainage systems and required an estimation of investment. This approach can also be applied to develop semi-hypothetical case studies to be used for research purposes.

The application of the algorithm to derive the drainage network layout in Belo Horizonte for the future scenario, shows that is promising, but its use still required supervision and expert knowledge to manually manipulate the data and transfer that knowledge to develop a model of the future drainage network. The shown limitations highlights the need to continue the research to polish the method, particularly to process the land use changes locally in every catchment to take care of the flow direction. Despite this limitation, the approach is valuable to test different scenarios of urban growth but also different approaches to deal with the urban runoff in the future.

7 Conclusions and Recommendations

7.1 Conclusions

7.1.1 Introduction

The aim of this research was to develop methods and tools that are capable of helping to design urban water systems for areas of urban expansion; and in particular to substantiate the vision of the urban planners and decision makers of the future urbanization in terms of appropriate water services. The exercise of visioning the future of the city is a way of expressing in words, images and intentions the future state of the city. Such visioning has to incorporate a number of different facilities and services, of which water is a key component. One of the best ways of exploring the implementation of appropriate water supply and drainage services for a new urban area is through the use of models. Models are accepted tools to help us explore hypotheses and to evaluate "what if" questions that test how the system will react or change to different parameters. This research has been a journey in exploring ways of transforming the formulation of visioning urbanization scenarios into models that contribute to our understanding of the future state of the water infrastructure in an urban area. The main drivers for understanding the future state are the dynamics of the urbanization and climate change. In this research we have focused on urban growth while neglecting climate change. Increasingly we live in an urbanizing world, but we have very little understanding about urban dynamics and how the interactions between many individuals and their decisions transform the city and affect its growth; the understanding of this process is still a challenge to address.

To give answers to the research questions presented in chapter 1 the conclusion has been grouped into sections.

7.1.2 Land use Change modeling

How to use the concepts of emergence and agent based techniques to urban water problems? In terms of water infrastructure our hypothesis was that by connecting the layout and magnitude of the water distribution and urban drainage networks to the distribution of land use in an urban area we can derive the future condition of the networks and therefore we can evaluate the impact of future developments on the existing systems. To be able to do this responsibly we need to know the possible future state(s) of the city. Several approaches reported in the literature were explored to explore models capable of doing this. Several ideas centered on the use of Agent-based models to reflect the urbanization process; in particular, we wanted to include the principles of emergence and the bottom -up approach in modeling. It was found that the cellular automata technique has been reported in the literature as promising, and it has already been applied to real cases in several cities with good success in terms of being capable of replicating the dynamics of urbanization and resembling what is observed in reality

To test the hypothesis different methods and algorithms were developed to be used in conjunction with several tools and existing models. These tools and methods were tested in two case studies: one for a water distribution network and the other one for an urban drainage network. The first case study used the water distribution network for the city of Villavicencio in Colombia, and the second case study adopted the sewerage system for the city of Birmingham in the UK.

For both case studies, data sets and the required information to build the land use change model were collected and processed. Moreover, in order to test the impact of the future development it was necessary to have a model of the corresponding network; this was one of the constraints in this research since models for water distribution and urban drainage networks are not always available or the authorities are unwilling to share them with the academia.

The first case study that was developed was for Villavicencio, Colombia. The initial model for urbanization was constructed using Metronamica. This is a cellular automata model developed by RIKS, in the Netherlands. Land use maps were available for Villavicencio until the year 1991 through the ILWIS datasets. The model was built to simulate the changes from 1978 to 1991, since this was the last map available. In addition to these maps, the elevation, slope and main road network were available to help explain the process of urban growth. The calibration process within this model consisted in adjusting manually the attraction/repulsion rules for every transition of land use and visually inspecting whether the result had the desired effect. The CA rules were based on weights acting along the central cell within the considered neighborhood. Due to unforeseen circumstances the cooperation with RIKS was not continued and so this model could not used anymore in this research. Despite the lack of data, a second model for land use change was built using DINAMICA EGO.

DINAMICA EGO is a freely available modeling tool developed in Brazil. It has been used in several research projects to investigate the impacts of land use change in urban areas and in the Amazon forest, etc. It was adopted in this research to provide the engine to construct the land use change models for the urban dynamics. Although the case study of Villavicencio was the first one that was developed, the case study of Birmingham offered a more complete data set. Therefore, the tools and methods developed for the land use change model were first tested with the information for Birmingham.

DINAMICA EGO allows the construction of the land use change model; nevertheless, the calibration procedures and the tools are not well described. For this reason, the computational engine was coupled with the NSGA II and NSGA XP to handle the calibration process as an optimization problem. Several intermediary algorithms were written to handle the optimization. The initial test was carried out using the Corine dataset for the years 1990 and 2000, and the fuzzy similarity test was used to assess the spatial correlation and to provide the objective function in the calibration process. In order to minimize computational demands the optimization loop was simulated with the NSGA II algorithm using parallel computing. The results show that the best option to make better use of the multiprocessor computer is to switch on the auto-detection number of processors included in Dinamica. Two approaches were evaluated to carry out the calibration of the model. The results obtained show that the two calibration approaches produce similar results. Nevertheless, the calibration approach that uses all of the parameters that deal with the process of expansion/contraction in the optimization loop is recommended. The step-by-step approach shows that a good correlation is found by an individual set of parameters at the beginning of the process; in subsequent steps a lower correlation can usually be achieved. The highest gain in correlation is achieved initially by adjusting the weight of evidence matrix that is at the heart of the land use transitions. Using this approach it was possible to adjust the model from 0.18 fitness value for the reciprocal similarity indicator to 0.46.

Despite the good results obtained initially, the visual inspection of the simulated land use map shows big discrepancies with the actual land use map of Birmingham for the year 2000. For this reason it was necessary to collect and analyse further data for this case study. It is important to highlight that even tough GA tools are good in handling the optimization process to adjust the model; remember, the purpose of the model is to replicate reality as close as possible. As in any model the accuracy of the simulations is directly related to the data being used. It is important that available information comes from the same source and that the methodology to generate the data remains consistent over time in order to make comparisons objective. In this research data that was freely available on the Internet was used. These data sets are becoming more reliable and are more accurate with time; still, care must be taken in analysing the data, especially data from different sources, taking into account the amount of data and the integration of the whole database. In this research it was evident that there is a lack of historical databases for urban infrastructure, land uses, water services and road expansions. This is important because understanding our past is important in identifying the causes and decisions that lead to the infrastructure we are using now, and this in turn has huge implications for the future.

After updating the land use change model for Birmingham, two models were generated with different cell resolutions. The calibration approach was repeated and applied to both models. The results showed that despite the results found initially, the adjustment of the Weights of Evidence did not produce a big gain in the fitness indicator. Due to the number of variables to be handled in the calibration process a sensitivity analysis was performed. The results of this analysis showed that the isometry factor in the function expander and patcher did not change across the different values that were assessed; therefore the factor was excluded from the optimization. This analysis was also useful to define the ranges of the parameters of the functions Expander and Patcher (mean patch size, variance patch size) better. There are very few data reported for thees type of parameters in the literature, perhaps because they differ from case to case, and , because this case attempts to simulate the internal dynamics of the urban area and not only the expansion of the urban core.

Is it possible to replicate the land use changes that are observed in reality by applying agent based methods? Are there differences between developed and developing countries? The results of the calibration showed that the updated land use change models were capable of replicating the final map considered as the objective for each model in general, the movements of land use, the expansion or generation of clusters, the spatial location, etc. The model M1 with a cell resolution of 30 meters achieved a fitness indicator of 62% and the model M2 with a cell resolution of 100 meter achieved a fitness indicator of 48%. This, together with the outcome of all the experiments conducted during the calibration of the models, showed that model M1 always scored higher than model M2. This indicates that a higher resolution has an influence and is important for the final result. But, it has to be mentioned that maintaining a high resolution in other datasets, although it may be useful to complement this analysis, is difficult and perhaps costly. This is particularly important for developing countries where this type of information and the necessary databases are more scarce and/or inaccurate as was found in this research with the case study in Colombia. Even though the fitness indicator is slightly lower, the resolution of 100 meters seems to be appropriate for scoping studies of urban planning and associated water infrastructure which can be used by municipalities and water companies. Different researchers suggest that this level of fitness is good for these types of studies. Especially if it is considered that there are intangible variables that cannot be measured and rules or political decisions that cannot be expressed in mathematical terms. The influence of these variables is such that they generate an impact that is perceived by the planners and developers at the local level. This also means that more effort needs to be put into developing this type of model to integrate it with others and to enhance it by incorporating knowledge from other areas in order to make them provide a more realistic simulation. Water services and their respective networks are socio-technical systems for which perhaps the social component is less well known; this also calls for a multidisciplinary approach.

7.1.3 Evolution of water distribution networks

This research is an attempt to develop an integrated approach for urban water analyses that can be used as planning tools to explore and evaluate future scenarios in urban areas. Traditionally this type of analysis is done using different tools in a fragmented way. The analysis is done with corridors along the main trunk pipe runs and the associated distribution of land use can be used to find rules to position the water mains for new areas of development. The developed approach that combines information from the land use and the road network looks promising; nevertheless it can drive the algorithm into areas that are not covered by the system. Further refinement is needed to improve the performance of the method.

Given a certain scenario of urban growth is it possible to identify the way to extend the water distribution network from the existing system to the new developments? The methods and tools developed and described here can be used to design the route of the water mains in expected new areas of development. Once the new system is connected to the existing system, the approach can be used to find rehabilitation measures that are required by the existing distribution network to maintain the level of service in the future.

7.1.4 Evolution of urban drainage networks

Given a certain future scenario of urban growth is it possible to identify the way to extend the urban drainage network from the existing system to the new developments? The methods presented and developed in this research for the drainage system in new urban areas demonstrate that connecting the distribution of land use in an urban area with other urban infrastructure information such as roads, canals, and drainage networks can help to derive key rules relating the different factors and to obtain a good approximation of the drainage network layout. Two techniques were used to derive the layout of the system, one using agent-based models and the other one using similar concepts to build a set of raster operations within ArcGIS. The approach was tested on a catchment in Birmingham, UK. The results show that both techniques perform well for a scoping analysis at the city level to derive the main pipes of the drainage network. As anticipated, the quality of the information affects both techniques. In particular, it was found that the cell size of the elevation map plays a major role.

In the case study area it was found that the main collectors of the wastewater pipes are located within 500 meters of the land use classes 'residential 1' (Continuous urban fabric) and 'industrial/commercial'. These are the main land use classes that influence the location of pipes in future scenarios.

The case study results show that the application of a cellular automata technique for simulating urban growth processes can yield promising results. The spatial analysis identified the need to model the internal changes in the city land use, and not only the expansion or contraction of the urban core. The development of the case study encountered several limitations regarding the nature and quality of the information needed to setup the model. The calibration process of the cellular automata model was undertaken despite the fact that not all the variables that are involved in the urbanization phenomena were used. The work performed highlights the need for a more multidisciplinary analysis.

To visualize the impacts of future urbanization growth on the existing drainage infrastructure, the output map of the cellular automata model for the year 2040 was used to derive a possible layout of the system in the future. The possible expansions of the drainage network were designed on the basis of the future land use map. The catchment parameters needed for the rainfall-runoff were estimated with the runoff extractor. The pipes were sized using the NSGA II algorithm coupled with SWWM within an optimization loop. The new pieces of the drainage network were then interconnected to the existing model of the drainage system to assess its new performance and to evaluate the consequences of the future land use changes for the existing infrastructure.

The interconnected model for the future urban growth scenario of Birmingham shows that the future developments will contribute further to the flooding problem if no improvements are made to the existing drainage system. The total number of flooded manholes will increase by 50 and most of the runoff volume that will be generated by the new developments will exacerbate the flood-related problems due to the lack of hydraulic capacity of the existing system. The approach presented in this paper can also be used to identify the critical pipes that require immediate attention for rehabilitation purposes. The same approach can be used to evaluate rehabilitation strategies to improve the performance of the system now and in the future.

There are still limitations of the methods presented here, the type of data required, the quality of the data and the availability of data affect the application of this approach. More cases need to be developed for further analysis. Information and data from climate change scenarios can be included to expand the capabilities of the analysis; the methods and algorithms are being further refined and tested.

7.2 Recommendations

The developed framework for modelling the water services infrastructure for new urban expansion areas was tested in two case studies taking into account land use change. It is recommended to test this approach in different case studies with different conditions of topography, climate and socio-economical development. With the use of information currently available on the Internet, the feasibility to use this methodology can be tested. This is particularly relevant for urban areas with limited information in terms of quantity and quality.

Some additional simulations should be carried out by changing the size and shape of the neighbourhood around the central cell in order to measure the influence of this parameter on the fitness of the output maps. Some of the possible neighbourhoods include: extended Moore neighbourhood, Circular Neighbourhood 8 Cells diameter, Rectangular Neighbourhood.

New simulations of land use change can also be undertaken. These can be done by generating new scenarios and changing the transition values from one land use to another. This would produce a different set of possible land use maps for the future by giving a wider range of possibilities into the future for city planners.

Changing the size of the neighbourhood when evaluating the fitness of the simulated landscape will help to find the best cell size to evaluate the performance of both models M1 and M2. In addition, the feedback of this size must be used to model the extent of the neighbourhood used in modelling the land use.

A better indicator of similarity between two different data sets of polylines could be developed. This should be done in order to take into consideration not only the closeness of the lines, but the alignments of such lines as well.

The generation of the expansion of the urban drainage layout only works with gravity systems and additional code must be written in order to be able to run the algorithm with systems that require pressurized systems and pumping stations.

In the creation of the cost weighted raster to generate the sewer network layout, equal importance was given to the information used to build this raster (Slope, Land use and roads). Different combination of weights associated with the various parameters should be tested to try to improve the performance of the generated layout.

Since the optimization performed within this research only considers the change in the pipe diameters further research should expand this analysis by including more parameters such as pipe slope and material.

Further research should also consider the design and rehabilitation of the existing drainage network where the new network is connected to it. This is needed to find which rehabilitation measurements are required by the existing infrastructure.

The inclusion of storage tanks and/or pumping stations in the optimization framework can improve the solution given by the optimization problem which currently focuses on changes in pipe diameter only.

References

Almeida, C.M.; Monteiro, A.M.V.; Câmara, G.; et al. (2005). Modelling Urban Land Use Dynamics through Bayesian Probabilistic Methods in a Cellular Automaton Environment. International Journal of Remote Sensing 26 (4). 759–774.

ASCE (2011). Failure to act. The economic impact of current investment trends in water and wastewater treatment infrastructure. USA.

Barredo, J.; Kasanko, M.; McCormick, N.; Lavalle, C. (2003). Modeling dynamic spatial processes: simulation of urban future scenarios through cellular automata. *Journal of Landscape and Urban Planning*. 64 (3). 145-160.

Barreto W (2012) Multi-Objective Optimization for Urban Drainage Rehabilitation., Ph. D dissertation. UNESCO-IHE, TU Delft. Delft, The Netherlands.

Barreto W, Vojinovic Z, Price RK, et al. (2006) Approaches to multi-objective multi-tier optimization in urban drainage planning. 7th international conference on hydro-informatics, Nice, Fr.

Batty, M. (1997). Cellular automata and urban form: a primer. *Journal of the American Planning Association* 63 (2). 266–274.

Batty, M. and Longley, P. (1994). *Fractal Cities. A geometry of form and function.* London: Academic Press.

Benenson, I. (1998). Multi-Agent simulations of residential dynamics in the city. *Computers, Environment and Urban Systems* 22 (1). 25-42.

Benenson, I. and Torrens, P. M. (2004). *Geosimulation: automata-based modeling of urban phenomena*. London: John Wiley.

Birmingham City Council (2012). Retrieved October, 2012, from www.birmingham.gov.uk.

Bonham-Carter G (1994) Geographic information systems for geoscientists: modelling with GIS. Pergamon, New York

Blumensaat, F.; Wolfram, M. and Krebs, P. (2011). Sewer model development under minimum data requirements. Environmental Earth Sciences 65 (5). 1427-1437.

Briassoulis, H. (2000). Analysis of Land Use Change: Theoretical and ModelingApproaches. Available Online. Retrieved January 2012 fromhttp://www.rri.wvu.edu/WebBook/Briassoulis/contents.htm.

Chen, Q. and Mynett, A.E. (2003). Effects of cell size and configuration in cellular automata based prey-predator modeling. *Simulation Modelling Practice and Theory* 11 (7-8). 609-625.

Clarke, K.; Hoppen, S. and Gaydos, L. (1997). A self-modifying cellular automaton model of historical urbanization in the San Francisco Bay area. *Environment and Planning B: Planning and Design* 24 (2). 247–261.

Dale, V.H. (1997). The relationship between land-use change and climate change.Ecological Applications (7). 753 – 769.

Darthe, B.; Sutherland, A.; Sharp, P.; Chlebek, J. and Mackay, R. (2008). A mid-term *review of switch in Birmingham, its learning alliance and engaging research to address the challenges of integrated water management.* Retrieved from SWITCH -http://www.switchurbanwater.eu/outputs/pdfs/W6-0_CBIR_REP_Review_Birmingham.pdf

De Leon, F.D.; Felsen, M. and Wilensky, U. (2007). *NetLogo Urban Suite - Tijuana Bordertowns model.* Retrieved from http://ccl.northwestern.edu/netlogo/models/UrbanSuite-TijuanaBordertowns. Center for Connected Learning and Computer-Based Modeling, Northwestern University, Evanston, IL

Diogo, A.F. and Graveto, V.M. (2006). Optimal Layout of Sewer Systems: A Deterministic versus a Stochastic Model. *Journal of Hydraulic Engineering* 132 (9). 927-943.

Ducrot, R.; Page, C.L.; Bommel, P. and Kuperb, M. (2004). Articulating land and water dynamics with urbanization: An attempt to model natural resources management at the urban edge. *Computers, Environment and Urban Systems* 28 (1-2). 85- 106.

Engelen, G.; Lavalle, C.; Barredo, J.I.; Van der Meulen, M. and White, R. (2007). *Modeling Land-Use Change, Progress and applications* 90. Book part V, 297-320. The Moland modeling framework for urban and regional land-use dynamics. Netherlands: Springer Netherlands.

Farolfi, S. and Hassan, R. M. (2003). AWARE: A Decision Support Tool towards decentralized water management in South Africa. International Colloquium in "Water Governance and Sustainable Development", Sousse (Tunisia), 9 & 10 October 2003.

Feuillette, S.; Bousquet, F. and Goulven, P. L.(2003). SINUSE: A multi-agent model to negotiate water demand management on a free access water table. *Environmental Modelling & Software* 18 (5). 413-427.

Fierro Patiño Manuel Javier. (2007). Factores que afectan la empresarialidad en el municipio de Villavicencio en el periodo comprendido entre los años 1994-2004. MSc Thesis. Universidad Pontificia Bolivariana, Medellin. Colombia.

Forrester W. J. (1968). Principle of systems. Wright-Allen Press.Cambridge,USA.

Forrester W. J. (1969). Urban Dynamics. PegasusCommunication. Waltham, USA.

FurtadoAlves Bernardo. (2009). Modeling social heterogeneity, neighborhoods andlocal influences on urban real estate prices. Spatial dynamic analyses in the Belo Horizonte metropolitan area, Brazil. PhD Dissertation. Faculty of Geoscience, University of Utrecht, The Netehrlands.

Geertman S., Hagoort M., and Ottens H. (2007). Spatial-Temporal specific neighborhood rules for cellular automata land use modelling. International journal of geographic information science. Vol 21, No. 5. 547-568.

Ghosh, I.; Hellweger, F.L.; Fritch, T.G. (2006). *Fractal generation of artificial sewer networks for hydrologic simulations*. Proceedings of the 26th Annual ESRI International User Conference, San Diego, CA.

Goodacre, C.M.; Bonham-Carter, G.F.; Agterberg, F.P.; Wright, D.F. (1993). A statistical analysis of spatial association of seismicity with drainage patterns and magnetic anomalies in western Quebec. *Tectonophysics* 217 (3-4). 285-305.

Guo Y., Walters G., Savic D. (2008a) Optimal design of storm sewer networks: Past, Present and Future. Proc., 11th Int. Conf. on Urban Drainage (ICUD 2008).

Guo Y., Walters G., Khu S.T., and Keedwell E. (2008b) Efficient Multiobjective Storm Sewer Design Using Cellular Automata and Genetic Algorithm Hybrid. Journal of Water Resources Planning and Management. Vol 134. No. 6. Pp 511-515.

Hagen, A. (2003). Fuzzy set approach to assessing similarity of categorical maps. *International Journal of Geographical Information Science* 17 (3). 235 – 249.

Haghighi, A. (2012). Loop by Loop Cutting Algorithm to generate Urban Drainage Systems Layout. *Journal of Water Resources Planning and Management*. 10.1061/(ASCE) WR.1943- 5452.0000294.

Hagoort, M. J. (2006). The neighbourhood rules: land-use interactions, urban dynamics and cellular automata modelling.Utrecht: Faculteit Geowetenschappen Universiteit Utrecht, v.334. 298 p. (Nederlands GeographicalStudies).

Hangouët, J.F. (1995). Computation of the Hausdorff distance between plane vector polylines. *Autocarto-Conference*. 1 – 10.

Hegde, N.P.; Muralikrishna, I.; Chalapatirao, K. (2005). Settlement growth prediction using neural network and cellular automata. *Journal of Theoretical and Applied Information Technology* 4 (5). 419 – 428

Houet, T. and Hubert-Moy, L. (2006). Modelling and projecting land-use and land-cover changes with a Cellular Automaton in considering landscape trajectories: An improvement for simulation of plausible future states. *EARSeL eProceedings* 5 (1). 63 – 76.

Kok, K. and van Delden, H. (2008). Combining two approaches of integrated scenario development to combat desertification in the Guadalentin watershed, Spain. *Environment and Planning B: Planning and Design* 36 (1). 49 – 66.

Koomen, E.; Stilwell, J.; Bakema, A.; Scholten, H.J. (2007). *Modelling Land-UseChange. Progress and Applications*. Springer.

Lambin, E.; Geist, H.; Lepers, E. (2003). Dynamics of land use and cover change in tropical regions. *Annual Review of Environment and Resources* (28). 205 – 241.

Last, E.W. (2011). *City water balance: a new scoping tool for integrated urban water management options*. Ph. D. Thesis. University of Birmingham.

Lechner, T.; Watson, B.A.; Tisue, S.; Wilensky, U.; Felsen, M. (2004). Procedural modeling of land use in cities. *Technical report NWU-CS-04-38.*

Lechner, T.; Watson, B.A.; Wilensky, U.; Felsen, M. (2003). *Proceduring city modeling.* 1st Midwestern Graphics Conference (St. Louis, MO).

Li, G. and Matthew, R.G.S. (1990). New approach for optimization of urban drainage system. *Journal of Environmental Engineering* 116 (5). 927–944.

Liu, Y. (2009). *Modelling Urban Development with Geographical Information Systems and Cellular Automata*. New York: Taylor & Francis Group.

Mair, M.; Sitzenfrei, R.; Möderl, M.; Rauch, W. (2012). Identifying multi utility network similarities. *World Environmental and Water Resources Congress 2012: Crossing Boundaries*. 3147 - 3153.

Medina, N. (2012). *Optimal design of sewer network systems under urbanization and land use change.* MSc Thesis. Delft: UNESCO-IHE.

Medina, N.; Sanchez, A.; Vojinovic, Z. (2012). Automatic Runoff Coefficient Estimation for Urban Drainage Modeling Using Google Maps Information and Fuzzy Classification. International Environmental Modelling and Software Society (iEMSs). International Congress on Environmental Modelling and Software Managing Resources of a Limited Planet, Sixth Biennial Meeting, Leipzig, Germany.

Meyer, W.B. and Turner II, B.L. (1994). Global Land-Use/Land-Cover Change.Cambridge: Cambridge University Press.

Moreno, N. and Marceau, D. (2007). Modeling land use changes using a novel vector-based geographic cellular automata. *MOAS'07 Proceedings of the 18th conference on Proceedings of the 18th IASTED International Conference: modelling and simulation*. 636 – 641.

Morton, D.; Rowland, C.; Wood, C. et al. (2011). Final report for LCM2007 – the new UK land cover map. *Countryside Survey Technical Report No 11/07*. 112.

Möderl, M.; Butler, D.; Rauch, W. (2009). A stochastic approach for automatic generation of urban drainage systems. *Water Science & Technology*. 59 (6). 1137–1143.

Price, R. K. (2004). Modelling Urban Drainage Systems: Part I Theoretical Development. *Hydro informatics Lecture Notes*. Delft: UNESCO-IHE.

RIKS. Retrieved January 2009, from *http://www.riks.nl*.

Sanabria, J., Hernández, T., Díaz, C, Gutierrez-García, H. (2008). A prototype GIS-based spatial model for analysing health service coverage and quality in Villavicencio's comuna 2. *Journal Orinoquia, Vol. 12, Núm. 2, pp. 197-210*. Universidad de Los Llanos

Sanchez, A. (2007). *Towards a demonstrator of an urban drainage decision support system*. MSc Thesis. Delft: UNESCO-IHE.

Sanchez, A.; Vojinovic, Z.; Price, R. (2012a). Exploring the changes in urban drainage networks based in cities future land use changes. *10th International Conference on Hydroinformatics*. Hamburg, Germany.

Sanchez, A.; Barreto, W.; Medina, N.; Vojinovic, Z. (2012b). Modeling Urban Growth and Land Use Change with Cellular Automata and Genetic Algorithms. International Environmental Modelling and Software Society (iEMSs). *International Congress on Environmental Modelling and Software Managing Resources of a Limited Planet, Sixth Biennial Meeting*. Leipzig, Germany.

Savic, D. and Walters, G. (1997). "Genetic Algorithms for Least-Cost Design of Water Distribution Networks." J. Water Resour. Plann. Manage., 123(2), 67–77.

Silva, E.A. and Clarke, K.C. (2002). Calibration of the SLEUTH urban growth model for Lisbon and Porto, Portugal. *Computers, Environment and Urban Systems* 26 (6). 525 – 552.

Sitzenfrei, R.; Fach, S.; Kinzel, H.; Rauch, W. (2010). A multi-layer cellular automata approach for algorithmic generation of virtual case studies: VIBe. *Journal of Water, Science and Technology* 61 (1). 37 – 45.

Soares-Filho, B.S.; Cerqueira, G.; Araujo, W.; Voll, E. (2011). DINAMICA project. Retrieved February 2011, from http://www.csr.ufmg.br/dinamica.

Soares-Filho, B.; Alencar, A.; Nepstad, D.; Cerqueira, G.; Vera Diaz, M.; Rivero, S.; Solórzano, L.; Voll, E. (2004). Simulating the Response of Land-Cover Changes to Road Paving and Governance Along a Major Amazon Highway: The Santarém-Cuiabá Corridor. *Global Change Biology* 10. 745 – 764. doi: 10.1111/j.1529-8817.2003.00769.x

Soares Filho, B.S.; Pennachin, C.L.; Cerqueira, G. (2002). DINAMICA – a stochastic cellular automata model designed to simulate the landscape dynamics in an Amazonian colonization frontier. *Ecological Modelling* 154 (3). 217 – 235.

Thuy Mai, T.T.P. (2009). *Integrated sustainability assessment of stormwater management strategies – a case study of The Catchment Upper Rea Main, Birmingham City, the United Kingdom*. MSc Thesis. Delft: UNESCO-IHE.

Torrens, P. M. (2003*). Automata-based models of urban systems*. In: Longley, P.A. and Batty, M., (eds.*)* Advanced spatial analysis: the CASA book of GIS**.** 61-80. Redlands: ESRI Press.

Tzima, F.; Athanasiadis, I. and Mitkas, P. (2006). *Report on the development of agent based models for water demand and supply*. Nostrum-DSS. EC.

UN-Habitat. (2008). States of the world's cities 2008-2009 / Harmonious cities. Earth Scan Publishing. London, UK.

Van Delden, H.; Luja, P. and Engelen, W. (2007). Integration of multi-scale dynamic spatial models of socio-economic and physical processes for river basin management. *Journal Environmental modeling and software* 22 (2). 223 – 238.

Van Vliet, J.; Hurkens, J.; White, R.; van Delden, H. (2012). An activity-based cellular automaton model to simulate land-use dynamics. *Environment and Planning B: Planning and Design* **39**(2). 198 – 212.

Verburg, P.H.; Schot, P.P.; Dijst, M.J. et al. (2004). *Land use change modelling:current practice and research priorities*. GeoJurnal (61). 309 – 324.

Villavicencio City Council (2012). Retrieved October, 2012, from www.villavicencio.gov.co.

Vojinovic, Z. (2005). Introduction to Urban Water Modelling. *Lectures notes in Hydro-informatics*. Delft: UNESCO-IHE.

Vojinovic, Z.; Sánchez Torres, A. and Barreto, W. (2008). Optimising Sewer System Rehabilitation Strategies between Flooding, Overflow Emissions and Investment Costs. *Paper presented at 11th International Conference on Urban Drainage.* Edinburgh, Scotland, UK.

Vojinovic, Z.; Solomatine, D.P. and Price, R.K., (2006). Dynamic Least-Cost Optimisation of Wastewater System Remedial Works Requirements. *Water Science and Technology* 54 (6-7). 467 – 475.

Von Neumann, J. (1949). *Theory of self-reproducing automata.* Urbana and London: University of Illinois Press.

Walski T.M., Chase D.V., Savic D.A., Grayman W.M., Beckwith S., Koelle E. (2003) *Advanced water distribution modeling and management.*Haestad Methods. USA

Watson, R. (2000). Summary for policy makers. Intergovernmental Panel on ClimateChange (IPCC). Special report on land use, land-use change, and forestry (LULUCF).Cambridge: Cambridge University Press.

White, R. and Engelen, G. (1994). Cellular dynamics and GIS: modelling spatial complexity. *Journal of Geographical Systems* 1. 237 – 53.

Wilensky, U. (2006). *NetLogo Grand Canyon model.* Retrieved from http://ccl.northwestern.edu/netlogo/models/GrandCanyon. Center for Connected Learning and Computer-Based Modeling, Northwestern University, Evanston, IL.

Wilensky, U. (1999). NetLogo. http://ccl.northwestern.edu/netlogo/. Center for Connected Learning and Computer-Based Modeling, Northwestern University, Evanston, IL.

Zellner, M.L. (2007). Generating policies for sustainable water use in complex scenarios: an agent-based model of Monroe County, Michigan. *Environment and Planning B: Planning and Design.* 34 (4). pp 664-686.

Zhang, Q.; Ban, Y.; Liu, J. et al. (2011.) Simulation and analysis of urban growth scenarios for the Greater Shanghai Area, China. *Computers, Environment and Urban Systems* 35 (2). 126 – 139.

Table of Figures

List of Tables

Appendix

TableA1 shows the original classification of the EAA data and its corresponding reclassification used in the present research.

TableA1. Original and reclassified classes of Model M2

Corine land cover land Use data (EEA)				Research's Reclassification	
Code	**Label 1**	**Label 2**	**Label 3**	**Code**	**Land Use**
1	Artificial surfaces	Urban fabric	Continuous urban fabric	2	Residential 1
2	Artificial surfaces	Urban fabric	Discontinuous urban fabric	3	Residential 2
3	Artificial surfaces	Industrial, commercial and transport units	Industrial or commercial units	4	Commercial/Industrial
4	Artificial surfaces	Industrial, commercial and transport units	Road and rail networks and associated land	0	Transportation
6	Artificial surfaces	Industrial, commercial and transport units	Airports	0	Airport
7	Artificial surfaces	Mine, dump and construction sites	Mineral extraction sites	0	Construction
8	Artificial surfaces	Mine, dump and construction sites	Dump sites	0	Construction
9	Artificial surfaces	Mine, dump and construction sites	Construction sites	0	Construction
10	Artificial surfaces	Artificial, non-agricultural vegetated areas	Green urban areas	0	Parks

Corine land cover land Use data (EEA)				Research's Reclassification	
Code	Label 1	Label 2	Label 3	Code	Land Use
11	Artificial surfaces	Artificial, non-agricultural vegetated areas	Sport and leisure facilities	5	Recreation
12	Agricultural areas	Arable land	Non-irrigated arable land	1	Vacant Land
16	Agricultural areas	Permanent crops	Fruit trees and berry plantations	1	Vacant Land
18	Agricultural areas	Pastures	Pastures	1	Vacant Land
20	Agricultural areas	Heterogeneous agricultural areas	Complex cultivation patterns	1	Vacant Land
21	Agricultural areas	Heterogeneous agricultural areas	Land principally occupied by agriculture, with significant areas of natural vegetation	1	Vacant Land
23	Forest and semi natural areas	Forests	Broad-leaved forest	1	Vacant Land
24	Forest and semi natural areas	Forests	Coniferous forest	1	Vacant Land
25	Forest and semi natural areas	Forests	Mixed forest	1	Vacant Land
26	Forest and semi natural areas	Scrub and/or herbaceous vegetation associations	Natural grasslands	1	Vacant Land
27	Forest and semi natural areas	Scrub and/or herbaceous vegetation associations	Moors and heathland	1	Vacant Land
29	Forest and semi natural areas	Scrub and/or herbaceous vegetation associations	Transitional woodland-shrub	1	Vacant Land

Corine land cover land Use data (EEA)				Research's Reclassification	
Code	**Label 1**	**Label 2**	**Label 3**	**Code**	**Land Use**
31	Forest and semi natural areas	Open spaces with little or no vegetation	Bare rocks	1	Vacant Land
41	Water bodies	Inland waters	Water bodies	0	Water

Table 3. Original and reclassified classes of Model M1, year 1990

Centre of Ecology and Hydrology Land use data		Research's Reclassification	
Code	**LCM Subclass Description**	**GRID_CODE**	**Land Use**
0	Unclassified	0	Not Modelled
1	Sea/Estuary	0	Not Modelled
2	Inland Water	0	Not Modelled
3	Beach and Coastal Bare	0	Not Modelled
4	Saltmarsh	0	Not Modelled
5	Grass Heath	1	Vacant Land
6	Mown / Grazed Turf	1	Vacant Land
7	Meadow / Verge Meadow / Verge / Semi-natural	1	Vacant Land
8	Rough / Marsh Grass	1	Vacant Land
9	Moorland Grass	1	Vacant Land
10	Open Shrub Moor	1	Vacant Land
11	Dense Shrub Moor	1	Vacant Land
12	Bracken	1	Vacant Land
13	Dense Shrub Heath	1	Vacant Land
14	Scrub / Orchard	1	Vacant Land
15	Deciduous Woodland	1	Vacant Land
16	Coniferous Woodland	1	Vacant Land
17	Upland Bog	0	Not Modelled
18	Tilled Land	1	Vacant Land
19	Ruderal Weed	1	Vacant Land
20	Suburban / Rural Development	3	Residential 2
21	Continuous Urban	2	Residential 1
22	Inland Bare Ground	1	Vacant Land
23	Felled Forest	1	Vacant Land
24	Lowland Bog	0	Not Modelled
25	Open Shrub Heath	1	Vacant Land

Table 4. Original and reclassified classes of Model M1, year 2000

Centre of Ecology and Hydrology Land use data		Research's Reclassification	
Code	**LCM Subclass Description**	**Code**	**Land Use**
221	Sea / Estuary	0	Not Modelled
131	Water (inland)	0	Not Modelled
201	Littoral rock	0	Not Modelled
211	Littoral sediment	0	Not Modelled
212	Saltmarsh	0	Not Modelled
181	Supra-littoral rock	0	Not Modelled
191	Supra-littoral sediment	0	Not Modelled
121	Bog (deep peat)	1	Vacant Land
101	Dense dwarf shrub heath	1	Vacant Land
102	Open dwarf shrub heath	1	Vacant Land
151	Montane habitats	1	Vacant Land
11	Broad-leaved / mixed woodland	1	Vacant Land
21	Coniferous woodland	1	Vacant Land
51	Improved grassland	1	Vacant Land
61	Neutral grassland	1	Vacant Land
52	Setaside grassland	1	Vacant Land
91	Bracken	1	Vacant Land
71	Calcareous grassland	1	Vacant Land
81	Acid grassland	1	Vacant Land
111	Fen, marsh, swamp	0	Not Modelled
41	Arable cereals	1	Vacant Land
42	Arable horticulture	1	Vacant Land
43	Arable non-rotational	1	Vacant Land
171	Suburban / rural developed	3	Residential 2
172	Continuous urban	2	Residential 1
161	Inland bare ground	1	Vacant Land

Table 5. Original and reclassified classes of Model M1, year 2007

Centre of Ecology and Hydrology Land use data			Research's Reclassification	
Code	**Broad habitat**	**LCM Subclass Description**	**Code**	**Land Use**
1	'Broadleaved, Mixed and Yew Woodland'	Broadleaved woodland	1	Vacant Land
2	'Coniferous Woodland'	'Coniferous Woodland'	1	Vacant Land
3	'Arable and Horticulture'	'Arable and Horticulture'	1	Vacant Land
4	'Improved Grassland'	'Improved Grassland'	1	Vacant Land
5		Rough Grassland	1	Vacant Land
6	'Neutral Grassland'	'Neutral Grassland'	1	Vacant Land
7	'Calcareous Grassland'	'Calcareous Grassland'	1	Vacant Land
8	'Acid Grassland'	Acid Grassland	1	Vacant Land
9	'Fen, Marsh and Swamp'	'Fen, Marsh and Swamp'	0	Not Modelled
10	'Dwarf Shrub Heath'	Heather	1	Vacant Land
11		Heather grassland	1	Vacant Land
12	'Bog'	'Bog'	1	Vacant Land
13	'Montane Habitats'	'Montane Habitats'	1	Vacant Land
14	'Inland Rock'	'Inland Rock'	0	Not Modelled
15	Salt water	Salt water	0	Not Modelled
16	Freshwater	Freshwater	0	Not Modelled
17	'Supra-littoral Rock'	'Supra-littoral Rock'	0	Not Modelled
18	'Supra-littoral Sediment'	'Supra-littoral Sediment'	0	Not Modelled
19	'Littoral Rock'	'Littoral Rock'	0	Not Modelled
20	'Littoral Sediment'	Littoral sediment	0	Not Modelled
21		Saltmarsh	0	Not Modelled
22	'Built-up Areas and Gardens'	Urban	2	Residential 1
23		Suburban	3	Residential 2

Annex 2. Landsat imagery for the years 1990, 2000 and 2006 for the area of Birmingham, UK.

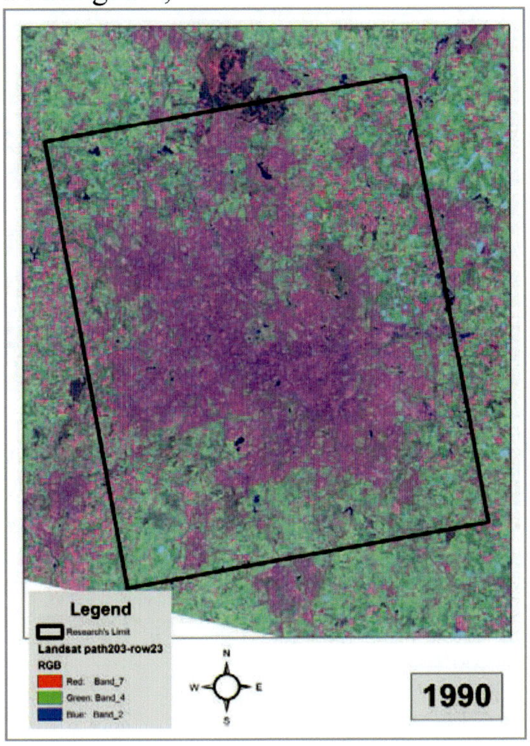

Figure 1. Landsat Image, 1990.

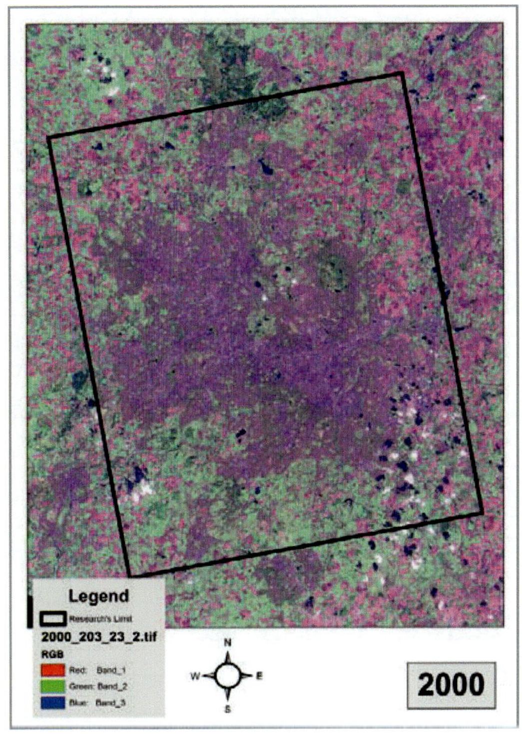

Figure 2. Landsat Image, 2000.

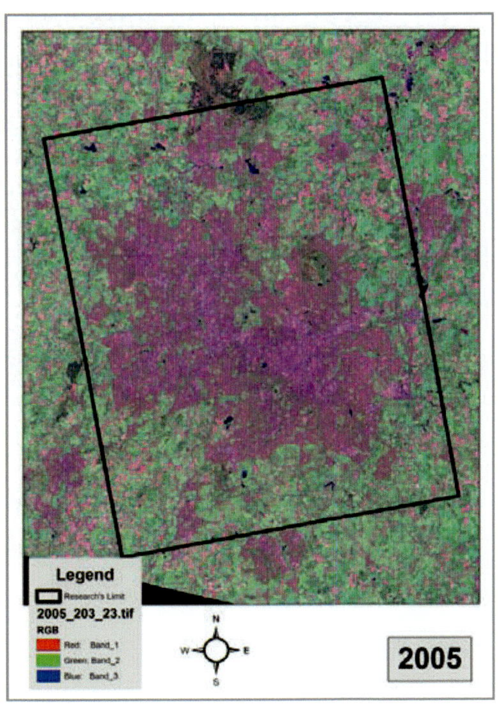

Figure 3. Landsat Image, 2006.

Annex 3. Table 0.1 and Table 0.4 for model M1 and from Table 0.5 to Table 0.8 for model M2.

Table 0.1. Modulate Change Matrix model M1. Calibrated Values

Single Step Modified		TO				
		1	2	3	4	5
FROM	1		0.729	0.060	0.844	0.724
	2	-----		0.990	0.012	0.494
	3	-----	0.781		0.834	0.291
	4	-----	0.598	0.414		0.438
	5	-----	0.331	0.812	0.729	

Table 0.2. Mean Patch Size Matrix model M1. Calibrated Values

Single Step Modified		TO				
		1	2	3	4	5
FROM	1		0.077	0.649	1.648	0.246
	2	-----		1.4157	1.299	0.919
	3	-----	0.001		1.159	1.959
	4	-----	1.696	0.761		1.352
	5	-----	1.007	0.895	1.511	

Table 0.3. Variance Patch Size Matrix model M1. Calibrated Values

Single Step Modified		TO				
		1	2	3	4	5
FROM	1		145.54	147.11	138.40	119.01
	2	-----		141.42	68.78	174.40
	3	-----	140.22		61.31	99.67
	4	-----	188.68	154.84		198.10
	5	-----	121.59	56.35	113.34	

Table 0.4. Variance Isometry model M1. Calibrated Values

Single Step Modified		TO				
		1	2	3	4	5
FROM	1		0.5	0.5	0.5	0.5
	2	-----		1.0	1.0	1.0
	3	-----	1.0		1.0	1.0
	4	-----	1.0	1.0		1.0
	5	-----	1.0	1.0	1.0	

Table 0.5. Modulate Change Matrix model M2. Calibrated Values

Single Step Modified		TO				
		1	2	3	4	5
FROM	1		-----	0.129	0.600	0.3223
	2	-----		0.7306	0.2449	-----
	3	-----	0.6955		0.1474	0.5084
	4	-----	-----	0.0875		0.7224
	5	-----	-----	0.7428	0.6011	

Table 0.6. Mean Patch Size Matrix model M2. Calibrated Values

Single Step Modified		TO				
		1	2	3	4	5
FROM	1		-----	1.65	1.14	1.86
	2	-----		1.96	1.12	-----
	3	-----	0.52		1.31	0.62
	4	-----	-----	1.35		0.35
	5	-----	-----	0.15	1.32	

Table 0.7. Variance Patch Size Matrix model M2. Calibrated Values

Single Step Modified		TO				
		1	2	3	4	5
FROM	1		-----	187.7	115.3	95.3
	2	-----		109.7	67.4	-----
	3	-----	164.9		156.0	142.5
	4	-----	-----	143.4		140.6
	5	-----	-----	165.6	138.4	

Table 0.8. Variance Isometry model M2. Calibrated Values

Single Step Modified		TO				
		1	2	3	4	5
FROM	1		0.5	0.5	0.5	0.5
	2	-----		1.0	1.0	1.0
	3	-----	1.0		1.0	1.0
	4	-----	1.0	1.0		1.0
	5	-----	1.0	1.0	1.0	

Annex 4. Area of Separation for all the simulated layouts for urban drainage network in Birmingham.

Exisitng network Vs Derived network Cost Weighted Raster 30 m.

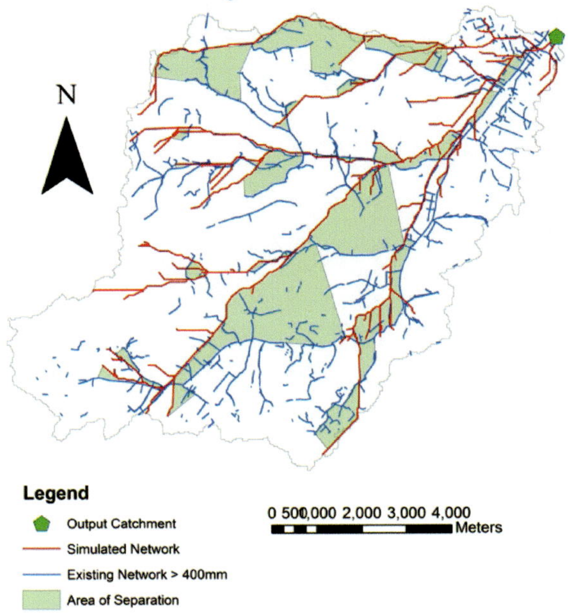

Exisitng network Vs Derived network Cost Weigthed raster 100 m.

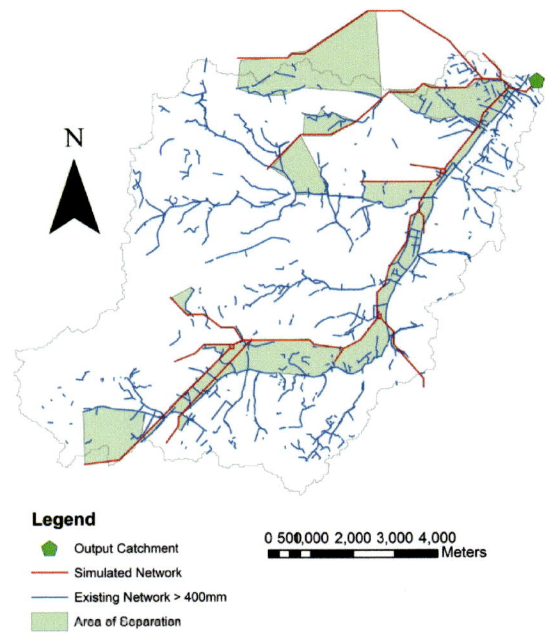

Exisitng network Vs Derived network Netlogo 30 m.

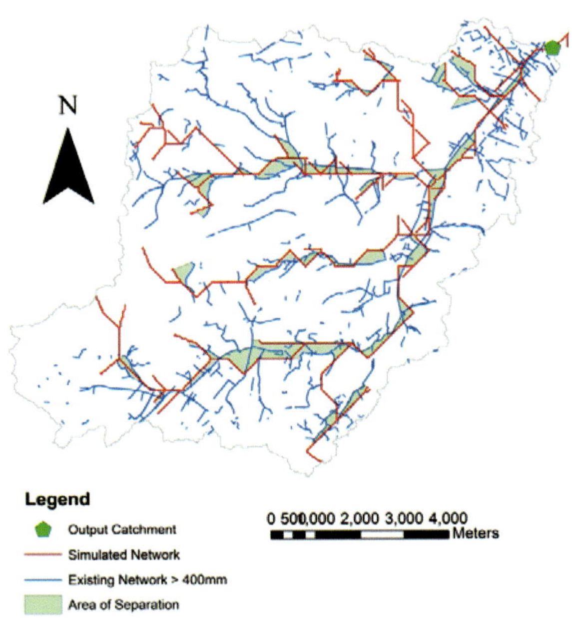

Exisitng network Vs Derived network Netlogo100 m.

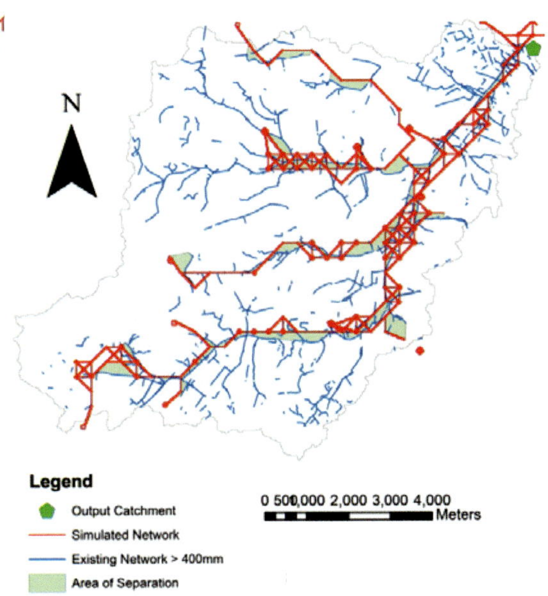

About the Author

Arlex Sanchez Torres was born in 1977 in Cali, Colombia. In 2001 he graduated as Sanitary Engineer form the Faculty of Engineering of the Universidad del Valle in Cali, Colombia. As an undergraduate he was already working at the university as student assistant for the sanitary chemistry laboratory. He was also student assistant for several projects at the Cinara Institute. After his graduation he joined Cinara where he was initially working in research projects to develop drinking water treatment technologies. By the end of 2001 he joined the International Water and Sanitation Centre based in Delft, the Netherlands as a junior professional officer for one and a half years. He developed skills to facilitate community management of water supply projects and enhanced his knowledge on water distribution modelling and demand management.

In 2003 he returned to Cinara where he continued to work in the field of water supply, water distribution networks modelling and efficient water use. In 2004 he started to work in the field of integrated water resources management at the catchment scale with the purpose of developing participatory planning approaches. At the time the vision of the environmental authorities was to enhance planning in the region. By participating in several projects the need to develop data acquisition tools and models to enhance decision making and planning, became evident. The knowledge gaps in his country as well as "love" directed him back to the Netherlands.

In 2005 he started his MSc studies in Hydroinformatics with a fellowship from an Alpha programme of the European Commission and the Unesco-IHE trust fund. He graduated in 2007 with a thesis that developed tools to optimize the rehabilitation of drainage networks considering costs, flood damages and water quality. After the birth of his daughter, he started his PhD studies in 2008 with the challenge of translating the vision of city planners for its future water resources into models that help them visualize the impact of their decisions.

His fields of interest are related to modelling the future water infrastructure of cities, in particular the connection of urban drainage network and water distribution networks with land use models of urban growth. This involves the collection of data and analysis of scenarios of future development, urbanization and land use change, together with the use and development of tools and methods to model future drainage infrastructures. At UNESCO-IHE, Arlex has been involved in several activities including lecturing, facilitation of groupwork and providing guidance to several MSc students.

List of Publications

Journal Papers

Sanchez A., Medina N., Vojinovic Z. and Price R. (2012). An integrated cellular automata evolutionary-based approach for evaluation of future scenarios and expansion of urban drainage networks. Paper Submmited to the Journal of Hydroinformatics. Under Revision.

Z. Vojinovic, S. Sahlu, A.S. Torres, S.D. Seyoum, F. Anvarifar, H. Matungulu, W. Barreto, D. Savic and Z. Kapelan.(2012). Multi-objective robust rehabilitation of urban drainage systems under uncertainties. Paper Submmited to the Journal of Hydroinformatics. Under Revision.

Sanchez A., Vojinovic Z., and Price R. K. (2013). Routing a water main in a new urbanising area. In preparation for the Journal of Hydroinformatics.

Sanchez A., Medina N., Mynett A. and Vojinovic Z. (2013). Modeling Urban Growth and Land Use Change with Cellular Automata and Genetic Algorithms , Case Study City of Birmingham, UK. In preparation for the Journal of Environmental Modeling and Software.

Conference Papers

Sanchez A., Vojinovic Z., and Price R. K. (2012). Exploring the changes in urban drainage networks based in cities future land use changes. 10th International Conference on Hydroinformatics. HIC 2012, Hamburg, Germany.

Medina N., Sanchez A. and Vojinovic Z. (2012). Automatic runoff coefficient estimation for urban drainage modeling using Google Maps Information and fuzzy classification. 6th International Congress on Environmental Modelling and Software - iEMSs 2012. Managing Resources of a Limited Planet: Pathways and Visions under Uncertainty. Germany.

Sanchez A., Medina N., Vojinovic Z. and Barreto W. (2012).Urban growth and Land Use Change Modeling, Using Cellular Automata and Genetic Algorithms. 6th International Congress on Environmental Modelling and Software - iEMSs 2012. Managing Resources of a Limited Planet: Pathways and Visions under Uncertainty. Germany.

Sanchez Torres A., Vojinovic Z., Price R. (2011). Determining the route for a water main in a new urbanising area. Proceedings of the Eleventh International Conference Computing and Control for the Water Industry: "Urban Water Management: Challenges and Opportunities". CCWI 2011. Exeter, UK.

Anvarifar F., Vojinovic Z., Sanchez Torres A. , Seyoum S. (2011). Use of evolutionary approaches for flood risk assessment and system rehabilitation. Proceedings of the Eleventh International Conference Computing and Control for the Water Industry: "Urban Water Management: Challenges and Opportunities". CCWI 2011. Exeter, UK.

Sanchez Torres A., Vojinovic Z., Price R and Waly M. (2011). Towards an Approach to the Evolution of Urban Drainage Networks Using Agent-Based Models. 12nd International Conference on Urban Drainage, Porto Alegre/Brazil, 11-16 September 2011.

Sanchez A., Vojinovic Z., and Price R. K. (2010). Planning urban water systems: modeling using cellular automata and numerical models. 9th International Conference on Hydroinformatics, HIC 2010. Tianjin, China.

Vojinovic Z., Matungulu H., Sanchez A., Seyoum S., and Barreto W., (2010). Multi-objective optimisation of urban drainage rehabilitation measures using evolutionary algorithms. 9th International Conference on Hydroinformatics, HIC 2010. Tianjin, China.

Vojinovic Zoran . and Sanchez Arlex. (2008). Optimising Sewer System Rehabilitation Strategies between Flooding, Overflow Emissions and Investment Costs, 11th International Conference on Urban Drainage, Edinburgh, Scotland.

Other Publications

Sanchez Arlex. 2007. Towards a demonstrator of an urban drainage decision support system. Master of Science Thesis. UNESCO-IHE. Delft, The Netherlands.

SANCHEZ Arlex, SANCHEZ Luis D. and VARGAS Silena. CD ROM. Multimedia Material on efficient use of water. 2004.

SANCHEZ Arlex, SMITS Stef, SANCHEZ Luis D. ; 2003. Recognizing reality ; multiple use of rural water supply systems. International Conference Agua 2003. Cartagena de Indias, Colombia.

SANCHEZ Arlex and SANCHEZ Luis Dario. 2003. Uso Eficiente de Agua. Thematic Overview Paper, TOP. IRC, International Water and Sanitation Centre; Instituto Cinara. Cali, Colombia.

SANCHEZ, Arlex. 2000. Comparación de Dos Configuraciones de Medio Filtrante en Clarificadores de Filtración Gruesa Ascendente en Capas. Tesis de Grado. Facultad de Ingenieria. Universidad del Valle. Cali, Colombia.

FERNÁNDEZ, Javier, SANCHEZ, Arlex; LATORRE, Jorge; MUÑOZ, Noel; BERÓN, Fabiola; y RESTREPO, Maribel. 2001. Uso de Fibras Naturales y Sintéticas en Filtración Lenta en Colombia. Una Experiencia en Ambientes Tropicales. Universidad del Valle – Cinara – Colciencias. Cali, Colombia.

For Product Safety Concerns and Information please contact our EU representative GPSR@taylorandfrancis.com Taylor & Francis Verlag GmbH, Kaufingerstraße 24, 80331 München, Germany

Batch number: 10722114

Printed by Printforce, the Netherlands